I0051251

Kushal Shah

Plasma and Plasmonics

Also of interest

Semiconductor Spintronics
Thomas Schäpers, 2016
ISBN 978-3-11-036167-4, e-ISBN: 978-3-11-042544-4

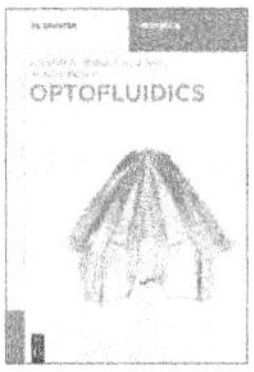

Optofluidics
Dominik G. Rabus, Cinzia Sada, Karsten Rebner, 2018
ISBN 978-3-11-054614-9, e-ISBN 978-3-11-054615-6

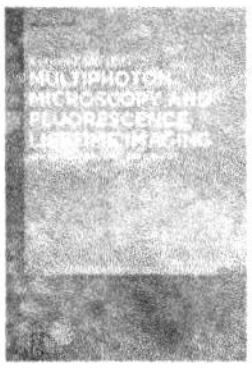

Multiphoton Microscopy and Fluorescence Lifetime Imaging
Applications in Biology and Medicine
Karsten König, 2017
ISBN 978-3-11-043898-7, e-ISBN 978-3-11-042998-5

Advances in Nanophotonics
Qihuang Gong, Zhi Li, Limin Tong, Yipei Wang, Yufei Wang, Wanhua Zheng, 2017
ISBN 978-3-11-030431-2, e-ISBN: 978-3-11-038288-4

Nanophotonics
Volker Sorger (Editor-in-Chief)
ISSN 2192-8614
http://nanophotonics-journal.com/

Kushal Shah

Plasma and Plasmonics

DE GRUYTER

Ane Books Pvt. Ltd.

Author
Kushal Shah
Department of Electrical Engineering and Computer Science,
Indian Institute of Science Education and Research (IISER),
Bhopal Bypass Road, Bhauri,
BHOPAL-462066, Madhya Pradesh, India

ISBN 978-3-11-056994-0
e-ISBN (E-BOOK) 978-3-11-057003-8
e-ISBN (EPUB) 978-3-11-057016-8

Library of Congress Cataloging-in-Publication Data
A CIP catalog record for this book has been applied for at the Library of Congress.

Bibliografische Information der Deutschen Nationalbibliothek
The Deutsche Nationalbibliothek lists this publication in the Deutsche Nationalbibliografie; detailed bibliographic data are available on the Internet at http://dnb.dnb.de.

Published by: Walter de Gruyter GmbH, Berlin/Boston
Under licence from: Ane Books Pvt. Ltd., New Delhi
Cover image: Equinox Graphics/Science Photo Library
Printing and binding: CPI books GmbH, Leck
♾ Printed on acid-free paper
Printed in Germany

www.degruyter.com

"The true logic of this world is the calculus of probabilities."

James Clerk Maxwell

Contents

Preface — ix

1 **Maxwell's Equations — 1**
1.1 Coulomb's and Gauss' Law — **2**
1.2 Faraday's Law — **4**
1.3 Ampere-Maxwell Law — **5**
1.4 Electromagnetic Waves in Free Space — **6**
1.5 Electromagnetic Waves in Matter — **9**
1.6 Snell's Law and Evanescent Waves — **13**
1.7 Group and Phase Velocity — **17**
1.8 Waveguides — **19**

2 **Electromagnetic Properties of Metals — 25**
2.1 Origin of Permittivity — **25**
2.2 Permittivity and Conductivity of Conductors — **27**
2.3 Electromagnetic Waves in a Conductor — **29**

3 **Plasma Kinetic Theory — 33**
3.1 Introduction — **33**
3.2 Exact Solutions for Time-Independent Electric Fields — **35**
3.3 Linear Response Theory (Plasma Waves) — **37**
3.4 Ponderomotive Theory — **39**
3.5 $\vec{E} \times \vec{B}$ Drift — **43**

4 **Plasma Fluid Theory — 45**
4.1 Introduction — **45**
4.2 Derivation of the Fluid Equations — **45**
4.3 Electrostatic Wave — **48**
4.4 Plasma Conductivity and Permittivity — **50**
4.5 Electromagnetic Waves — **51**

5 **Surface Plasmon Polaritons (SPP) — 55**
5.1 SPP on Single Interface — **56**
5.1.1 TE Mode of SPP — **57**
5.1.2 TM Mode of SPP — **58**

5.2 SPP on Multilayer Systems — 60
5.3 Excitation of SPP — 62
5.4 Localized Surface Plasmon Resonance (LSPR) — 65
5.5 Aplications of Surface Plasmons — 69

6 Spoof Surface Plasmons (SSP) — 71
6.1 SSP at Low Frequencies — 71
6.2 SSP at High Frequencies — 73
6.3 Self-collimation in SSP — 77

7 Advanced Topics in Plasmonics — 83
7.1 Negative Index Metamaterials (NIMs) — 83
7.2 Surface-Enhanced Raman Scattering (SERS) — 88
7.3 Particle Traps — 90

8 Mathematical Foundations — 99
8.1 Scalars and Vectors — 99
8.1.1 Coordinate Systems: Cartesian, Cylindrical and Spherical — 99
8.1.2 Gradient of a Scalar — 104
8.1.3 Divergence and Curl of a Vector — 105
8.1.4 Scalar and Vector Integration — 107
8.1.5 Vector Identities — 110
8.1.6 Scalar and Vector Potential — 111
8.2 Lorentz Transformations and Special Relativity — 113
8.3 LTI Systems and Green's Function — 117
8.4 Fourier Transform — 121
8.5 Linear Stability Analysis of ODEs — 125
8.6 Hamiltonian Formulation of Charged Particle Dynamics — 134

9 Numerical Methods for Electromagnetics — 139
9.1 Laplace Equation — 139
9.2 Runge-Kutta Method — 141
9.3 Wave Equation: FDTD Method — 144
9.4 FDTD Dispersion Relation — 146
9.5 Dispersive Materials — 148

Appendix: Legendre Polynomials — 153

Bibliography — 155

Index — 157

Preface

Electromagnetics is one of the most fundamental and unique phenomenon of nature. Electromagnetic fields are governed by the Maxwell's equations which were first conceived by James C. Maxwell in the early 1860s. Various components of the Maxwell's equations were formulated by scientists like Carl F. Gauss, Michael Faraday and Andre-Marie Ampere. Maxwell's main contribution was to combine these various disconnected equations and show that they are deeply related to each other. Maxwell's original formulation was actually quite complicated and it was simplified into the vector calculus formulation that is commonly used by Oliver Heaviside in the 1880s.

One of the most fundamental predictions that emerged as a result of combining the various electric and magnetic field equations was that of electromagnetic waves. These waves carry light and energy from the sun to our earth and thus, are fundamental to the existence of life on this planet. They also set the maximum speed that is possible to attain by any object/wave in the physical world and hence, form the basis for the theoretical modeling of all the other fundamental forces of nature. Thus, Maxwell's equations are in some way central to both physics and biology.

There are many kinds of electromagnetic waves which have been studied over the past 100 years and more. In this book, we focus on a particular kind of electromagnetic wave that travels on the interface between a metal and a dielectric. These waves, known as surface plasmons, were discovered a few decades back and have found very interesting applications in communications and design of biological and chemical sensors. In order to understand these waves, it is also important to understand how electromagnetic waves interact with charged particles (plasmas). This is because, in the frequency regime where we have plasmon waves, the electrons in a metal essentially behave like a plasma and mimic many of its properties.

There are several well written books on electromagnetic fields, plasmas and plasmonics. And I will be always indebted to David J. Griffiths, John D. Jackson, Dwight R. Nicholson, Stefan Maier and others for writing such wonderful books which generated within me a very deep interest in this subject. In this book, I would like to combine various aspects from these three topics and present sufficient material that can form a semester long course at the postgraduate or senior undergraduate level.

I would like to specially thank Chapala Mohanty, my teacher during school days, for introducing me to the beautiful world of mathematics and giving me the wings to fly higher than what I thought I could. I am grateful to Anil Prabhakar, Sudip Sengupta and Rajaraman Ganesh who taught me many concepts of electromagnetic fields and plasma physics. I have no words to express my indebtedness to Harishankar Ramachandran, my PhD advisor, who not only introduced me to the wonderful world of plasma physics but also taught me to question every single scientific concept and to put more emphasis on truth rather than on authority. I also thank my first PhD student, Sayak Bhattacharya, who introduced me to the exciting field of plasmonics. I hope

https://doi.org/10.1515/9783110570038-009

this book inspires many other students to further explore the open problems in these areas.

Lastly and most importantly, I would like to thank my parents who gave me the freedom to pursue my heart's desire and the strength to go beyond my own limitations. I am also immensely grateful to my wife, Shweta Arya, for being so supportive in all my endeavors and for bringing in the much needed stability in my life.

Kushal Shah

New Delhi, 06 March 2017

1 Maxwell's Equations

Scientists and philosophers are yet to provide a satisfactory answer to what constitutes *life* and what differentiates the world of the living from the rest. But one thing that is undisputed is that life would not exist without *light*. The rest would. Some philosophers have even gone to the extent of saying that light is in fact the very essence of life. Interestingly, although light may not be really essential to the existence of non-living things, it has been of crucial importance in physics which is a study of the very fundamental aspects of non-living objects. Is it just a coincidence? Almost all the discoveries of modern physics (both experimental and theoretical) in the 20th century and later have depended to a very large extent on our understanding of light. Even the recent discovery of gravitational waves would not have been possible without our mastery of precision laser systems. So by all means, let there be light!

Light is essentially a kind of wave that propagates in free space. Its most crucial property is that it doesn't need any medium for its existence or propagation. This is also perhaps why it is so fundamental to everything else in the universe. Now this is common knowledge but back in the 19th century, the idea of a wave traveling without a medium was very hard to believe. It took a very serious effort by two scientists [Albert A. Michelson and Edward W. Morley] to get over this limitation of thought. The wave of light is made up of two components: electric field and magnetic field. Together these are known as the *electromagnetic fields*. As the light wave propagates in free space, the electric field and magnetic field keep changing their strength periodically but always remain mutually perpendicular to each other and also perpendicular to the direction of propagation. This condition of perpendicularity may not be true when the light wave propagates in certain media. In order to understand the nature of the light wave and its constituent electromagnetic fields, scientists needed an equation. As it turned out, one equation was not sufficient but four did the job. This set of four equations describing light wave or any other kind of electromagnetic phenomenon is known as the **Maxwell's equations**,

$$
\begin{array}{ll}
\vec{\nabla} \cdot \vec{E} = \dfrac{\rho}{\epsilon_0} & \text{Gauss' Law} \\
\vec{\nabla} \cdot \vec{B} = 0 & \text{Gauss' Law for Magnetism} \\
\vec{\nabla} \times \vec{E} = -\dfrac{\partial \vec{B}}{\partial t} & \text{Faraday's Law} \\
\vec{\nabla} \times \vec{B} = \mu_0 \vec{J} + \mu_0 \epsilon_0 \dfrac{\partial \vec{E}}{\partial t} & \text{Ampere-Maxwell Law}
\end{array}
\qquad (1.1)
$$

We will see details of these equations and the meaning of these symbols in later sections of this chapter. And the rest of the book is dedicated to understanding the interesting properties of these light waves as they propagate through plasmas and metals.

https://doi.org/10.1515/9783110570038-011

Maxwell's equations are one of the most important set of equations in all of physics. It is these equations and their related concepts that have led to the birth of a large part of modern physics ranging from special relativity to quantum mechanics. These equations also hold a unique distinction. Almost all the equations in physics that were known before the beginning of 20th century were later found to be approximations to a much deeper reality and thus have undergone a change in one way or the other. Maxwell's equations are perhaps the only ones (or one of the very few) to have survived the deepest scrutiny. Numerous experiments have been performed to understand the nature of light but none of them found the slightest need to rethink about these equations. And lets hope that it remains like this! If a correction is ever required in Maxwell's equations, we may have to rethink about all our cherished concepts in modern physics! Now of course it must be noted that Maxwell's equations only deals with the wave nature of light and does not say anything about the particle nature (photons). But even the equations describing the behavior of photons (quantum field theory) uses the Lorentz transformation that is an integral part of Maxwell's equations.

1.1 Coulomb's and Gauss' Law

That matter has electrical properties has been known to mankind since ancient times. But it was only in the 17th century that a serious study of these properties was undertaken. Several scientists did experiments that seemed to show an inverse-square law kind of behavior for the force between two charged objects. However, it was only in 1785 that *Charles-Augustin de Coulomb* convincingly showed that the force between to charged particles obeys the law that now bears his name,

$$F = k\frac{q_1 q_2}{r^2} \tag{1.2}$$

where $k = 8.99 \times 10^9$ N m^2 C^{-2} is the Coulomb's constant, q_1, q_2 are the charges on the particles and r is the distance between them. If there are multiple charged particles present, then the total force on a particle of charge q_i and position $\vec{r}_i$ is given by the

Tab. 1.1: Timeline of the discovery of various components of Maxwell's equations.

Year	Law	Country
1785	Coulomb's law	France
1813	Gauss' law	Germany
1823	Ampere's law	France
1831	Faraday's law	England
1861	Maxwell's equations	England (Scotland)

vector sum of all the forces individually exerted by all other particles,

$$\vec{F}_i = kq_i \sum_{j=1, j \neq i}^{N} \frac{q_j}{\left|\vec{r}_i - \vec{r}_j\right|^3} \left(\vec{r}_i - \vec{r}_j\right) \tag{1.3}$$

If instead of having N individual particles, we have an extended charged body, the force on the particle can be found by taking an integral instead of the summation in Eq. (1.3). However, for extended bodies, it is more convenient to write the above law in terms of charge densities instead of actual charge. This is because in an extended body, the total charge at any given point is zero and its only an infinitesimal length/area/volume that has a non-zero charge. For point particles, the charge density is simply given by a Dirac delta function with an appropriate scaling.

There is another equivalent form of writing Coulomb's law above by using the electric flux instead of the electric force. As we will see later, this form known as the Gauss' law (named after Johann Carl Friedrich Gauss) is much more useful while dealing with electromagnetic waves. Johann Carl Friedrich Gauss was a German mathematician and has contributed immensely to many other fields in mathematics and is ranked as one of history's most influential mathematicians.

If we have a point particle of charge q at the origin of a 3D coordinate system, the electric field at any other point in space $(\vec{r})$ is given by

$$\vec{E} = \frac{kq}{r^2}\hat{r} \tag{1.4}$$

The electric flux going through the surface of a sphere of radius r centered around the origin is given by

$$\Phi_E = \oint \vec{E} \cdot \vec{dA} \tag{1.5}$$

where $\vec{dA}$ is a vector representing an infinitesimal area on the sphere surface at location $\vec{r}$ and $(\cdot)$ represents the dot product. The direction of $\vec{A}$ is the outward normal to the surface at point $\vec{r}$. Using spherical polar coordinates, we get $\vec{dA} = r^2 \sin\theta d\theta d\phi \hat{r}$, where $\theta \in [0, \pi]$ is the inclination angle and $\phi \in [0, 2\pi]$ is the azimuthal angle. Substituting this expression for $\vec{dA}$ and Eq. (1.4) in Eq. (1.5), we get

$$\begin{aligned}\Phi_E &= kq \oint \sin\theta d\theta d\phi \\ &= 4\pi kq\end{aligned} \tag{1.6}$$

Though we have derived the above equation assuming a point charge and a symmetrically placed spherical sphere, the same result is obtained for any arbitrary charge configuration and any arbitrary closed surface. Writing $k = 1/(4\pi\epsilon_0)$, we get the integral form of the *Gauss' law*,

$$\oint \vec{E} \cdot \vec{dA} = \frac{Q}{\epsilon_0} \tag{1.7}$$

By using the divergence theorem, this can be converted to the differential form which we are going to use in this book,

$$\vec{\nabla} \cdot \vec{E} = \frac{\rho}{\epsilon_0} \tag{1.8}$$

where ρ is the charge density.

For the case of magnetic fields, there is an equation equivalent to that of electric fields and is known as the *Gauss' law of magnetism*,

$$\vec{\nabla} \cdot \vec{B} = 0 \tag{1.9}$$

Here the right hand side is zero since magnetic poles have been always found to appear in pairs and a region of space has never been found to have a net magnetic charge. Though the existence of these magnetic monopoles has been predicted by some version of particle theories (e. g., superstring theory), so far all efforts to find magnetic monopoles have been unsuccessful. In 1931, the Nobel laureate Paul A. M. Dirac showed that if there are any magnetic monopoles present anywhere in the universe, then all the electric charge present in the universe must be quantized. We do know that electric charge is quantized at the microscopic level, but this still does not prove the existence of magnetic monopoles. If a magnetic monopole is indeed found in experiments, it would lead to a change in Eq. (1.9) and perhaps many ideas that we currently hold dear might have to be given up.

1.2 Faraday's Law

Michael Faraday has been one of the pillars of electromagnetics since he not only discovered the induction law that bears his name but also predicted the existence of electromagnetic waves that have now become the bedrock of science as well as our daily lives. In a series of experiments conducted in the 1820s, Faraday discovered that a change of magnetic flux through a closed conducting wire loop leads to a flow of current. This Faraday's law is usually stated as, *"The induced electromotive force in any closed circuit is equal to the negative of the time rate of change of the magnetic flux enclosed by the circuit"*. Faraday was not a mathematician and so he didn't present his ideas in a more quantitative fashion. That job was done by Maxwell and the equation he came up with is,

$$\vec{\nabla} \times \vec{E} = -\frac{\partial \vec{B}}{\partial t} \tag{1.10}$$

which is the divergence form of the Faraday's law. Using the Kelvin-Stokes' theorem, this can also be written in the integral form,

$$\oint \vec{E} \cdot \vec{dl} = -\frac{\partial \Phi_B}{\partial t} \tag{1.11}$$

where the integral is taken along a closed loop and Φ_B is the magnetic flux (magnetic field times the area) passing through the area enclosed by the loop. Here it is important to note that the electric field induced as a result of changing magnetic flux is non-conservative, which means that its integral around the closed loop is non-zero. This means that this electric field cannot be written as the derivative of a scalar potential as can be done in electrostatics. We will discuss more about these potentials in a later section of this chapter.

1.3 Ampere-Maxwell Law

Electric and magnetic fields are deeply interconnected. By Faraday's law, we know that a changing magnetic field produces an electric field. It is natural to ask if the reverse is also true. And the answer is certainly that yes, a changing electric field also produces a magnetic field. This is described by what is known as the Ampere-Maxwell law. However, there is a very important difference between the electric and magnetic field in this context. From the integral form of Faraday's law (Eq. (1.11)), we know that the line integral of the electric field around a closed loop is zero if there is no time-varying flux passing through the area enclosed by the loop. But the line integral of a magnetic field around a closed is non-zero even if there is a constant current passing through the area enclosed. In the absence of time-varying electric fields, this magnetic field line integral is given by what is known as the *Ampere's circuital law*,

$$\oint \vec{B} \cdot d\vec{l} = \mu_0 I_{encl} \tag{1.12}$$

where μ_0 is the permeability of free space and I_{encl} is the total current passing through the area of the loop. This can also be written in the differential form,

$$\vec{\nabla} \times \vec{B} = \mu_0 \vec{J} \tag{1.13}$$

where $\vec{J}$ is the current density. So we now have four equations governing the behavior of electromagnetic fields: Gauss' law for electric fields (Eq. (1.8)), Gauss' law for magnetic fields (Eq. (1.9), Faraday's law (Eq. (1.10)) and Ampere's circuital law (Eq. (1.13)). Each of these equations captures a certain aspect of the electromagnetic waves. For these equations to jointly describe electromagnetic fields, it is necessary for them to be consistent with each other. This implies that none of the equations should contradict the others. What Maxwell discovered was that these equations do contradict each other and need to be slightly modified in order to ensure consistency. Now this must also be done without the resulting equation disputing already known experimental results. As it turned out, a simple modification of the Ampere's circuital law (Eq. (1.13)) did the trick and this also led to the mathematical prediction of electromagnetic waves which Faraday had intuitively predicted long ago.

In order to discuss the inconsistency in the Ampere's circuital law, we must first introduce the *continuity equation* which directly follows from conservation of charge

$$\vec{\nabla} \cdot \vec{J} = -\frac{\partial \rho}{\partial t} \tag{1.14}$$

This equation says that if in a region of space the charge density is changing, then the change must be carried away/into by a current. This must obviously be true if the total charge is to be conserved. In classical mechanics, charge can only be transported from one location to another but never be created or destroyed. Now if we take a divergence of the Ampere's circuital law (Eq. (1.13)), the left hand side becomes zero (since the divergence of a curl is always zero), but the right hand side may not be zero if the charge density is changing with time (Eq. (1.14)). In order to fix this problem, let us substitute the Gauss' law (Eq. (1.8)) into the continuity equation (Eq. (1.14)) to get,

$$\vec{\nabla} \cdot \vec{J} = -\epsilon_0 \frac{\partial \vec{\nabla} \cdot \vec{E}}{\partial t}$$

$$\Rightarrow \vec{\nabla} \cdot \left(\vec{J} + \epsilon_0 \frac{\partial \vec{E}}{\partial t} \right) = 0$$

Thus, it follows that although the divergence of the current density may not be zero, the divergence of the quantity in brackets in the above equation will always be zero. Maxwell conjectured that the correct form of Eq. (1.13) must be

$$\vec{\nabla} \times \vec{B} = \mu_0 \vec{J} + \mu_0 \epsilon_0 \frac{\partial \vec{E}}{\partial t} \tag{1.15}$$

where the term $\epsilon_0 \partial \vec{E} / \partial t$ is known as the displacement current. This is now known as the Ampere-Maxwell law.

Thus, we now have the complete set of equations needed to described the electromagnetic waves and all together these are known as the Maxwell's equations,

$$\vec{\nabla} \cdot \vec{E} = \frac{\rho}{\epsilon_0} \qquad \text{Gauss' Law}$$

$$\vec{\nabla} \cdot \vec{B} = 0 \qquad \text{Gauss' Law for Magnetism}$$

$$\vec{\nabla} \times \vec{E} = -\frac{\partial \vec{B}}{\partial t} \qquad \text{Faraday's Law}$$

$$\vec{\nabla} \times \vec{B} = \mu_0 \vec{J} + \mu_0 \epsilon_0 \frac{\partial \vec{E}}{\partial t} \qquad \text{Ampere-Maxwell Law}$$

1.4 Electromagnetic Waves in Free Space

The simplest and yet the most powerful implication of the Maxwell's equations is the existence of plane waves which propagate at a constant speed in free space. This speed know as the *speed of light* is one of the corner stones of Einstein's theory of relativity and many other aspects of modern physics. In order to derive the expression for these waves, we first set the charge density and current density to zero $\left(\rho = 0 = \vec{J}\right)$ since we are interested in free space. In the next section we will discuss the properties of these waves in a material medium. Thus, for free space, the Maxwell's equations become

$$\begin{aligned}\vec{\nabla} \cdot \vec{E} &= 0 \\ \vec{\nabla} \cdot \vec{B} &= 0 \\ \vec{\nabla} \times \vec{E} &= -\frac{\partial \vec{B}}{\partial t} \\ \vec{\nabla} \times \vec{B} &= \mu_0 \epsilon_0 \frac{\partial \vec{E}}{\partial t}\end{aligned} \tag{1.16}$$

Taking a curl of the third equation in the above set and substituting for $\vec{\nabla} \times \vec{B}$ from the fourth equation, we get

$$\begin{aligned}\vec{\nabla} \times \left(\vec{\nabla} \times \vec{E}\right) &= -\mu_0 \epsilon_0 \frac{\partial^2 \vec{E}}{\partial t^2} \\ \Rightarrow \vec{\nabla}\left(\vec{\nabla} \cdot \vec{E}\right) - \nabla^2 \vec{E} &= -\mu_0 \epsilon_0 \frac{\partial^2 \vec{E}}{\partial t^2} \\ \Rightarrow \nabla^2 \vec{E} - \mu_0 \epsilon_0 \frac{\partial^2 \vec{E}}{\partial t^2} &= 0\end{aligned} \tag{1.17}$$

which is the equation of a plane electromagnetic wave in free space and $c = \sqrt{1/\mu_0 \epsilon_0} = 3 \times 10^8\ \mathrm{m/s}$ is the speed of light in vacuum. Note that in order to get the last step in Eq. (1.17), we have used the Poisson equation in charge free space, $\vec{\nabla} \cdot \vec{E} = 0$.

Now we need to solve Eq. (1.17) in order to obtain the expression of the electric field comprising the electromagnetic wave. In order to do this, we take a Fourier transform of this equation, to get

$$-k^2 \vec{E}\left(\vec{k}, \omega\right) + \mu_0 \epsilon_0 \omega^2 \vec{E}\left(\vec{k}, \omega\right) = 0$$

which implies that $\vec{E}\left(\vec{k}, \omega\right)$ can be non-zero only when

$$\omega^2 = c^2 k^2 \tag{1.18}$$

which is the dispersion relation of a plane electromagnetic wave in vacuum. Thus, the general expression for an electric field that satisfies Eq. (1.17) is given by

$$\vec{E}\left(\vec{r}, t\right) = \iiint \vec{A}\left(\vec{k}\right) \exp\left[i\left(\vec{k} \cdot \hat{r} - ckt + \phi_k\right)\right] \mathrm{d}^3 k \tag{1.19}$$

where $\vec{A}\left(\vec{k}\right)$ is an arbitrary vector that satisfies $\vec{k} \cdot \vec{A} = 0$ (this follows from Poisson's equation in free space, $\vec{\nabla} \cdot \vec{E} = 0$), the triple integral represents integration over each of the 3 components of $\vec{k}$ and ϕ_k is the phase and depends on k. The expression for the electric field in Eq. (1.19) represents a superposition of many plane waves traveling along all possible directions. This superposition is a very important feature of plane waves and follows from the linearity of Maxwell's equations. This property also holds true in all kinds of linear media, which we will discuss in the next section. There are also many important kinds of nonlinear media where the electric field cannot be written as a superposition of individual Fourier components. However, we will not be discussing these nonlinear effects in this book.

A plane wave refers to one of the components of this general expression (Eq. (1.19)) and is given by

$$\vec{E}\left(\vec{r}, t\right) = \vec{E}_0 \exp\left[i\left(\vec{k}_0 \cdot \hat{r} - ck_0 t + \phi_0\right)\right] \tag{1.20}$$

where $\vec{k}_0 \cdot \vec{E}_0 = 0$ and $\vec{k}_0$ is the direction of propagation of the wave. This wave is called a plane wave since its wavefront is a plane surface. Substituting the above expression for $\vec{E}$ in the Faraday's law, we can now find the expression for the magnetic field

$$\vec{B}\left(\vec{r}, t\right) = \frac{\left(\vec{k}_0 \times \vec{E}_0\right)}{ck_0} \exp\left[i\left(\vec{k}_0 \cdot \hat{r} - ck_0 t + \phi_0\right)\right] \tag{1.21}$$

It can be clearly seen from Eqs. (1.20) and (1.21) that for a plane electromagnetic wave, the electric field vector $\left(\vec{E}\right)$, the magnetic field vector $\left(\vec{B}\right)$ and the direction of propagation $\left(\hat{k}\right)$ are all perpendicular to each other. Writing Eq. (1.21) only in terms of the unit vectors, we get

$$\hat{B} = \hat{k} \times \hat{E}$$

Taking a cross product of $\hat{E}$ with each side of the above equation, we get

$$\begin{aligned} \hat{E} \times \hat{B} &= \hat{E} \times \left(\hat{k} \times \hat{E}\right) \\ &= \hat{k}\left(\hat{E} \cdot E\right) - \hat{E}\left(\hat{E} \cdot \hat{k}\right) \\ \Rightarrow \hat{k} &= \hat{E} \times \hat{B} \end{aligned} \tag{1.22}$$

since $\hat{k} \cdot \hat{E} = 0$ for plane waves in vacuum.

Here it is important to note that plane waves are a mathematical idealization and strictly speaking, do not actually exist in reality. This is because in order to generate a perfect plane wave, we need an infinitely large conducting surface with sinusoidally oscillating currents. In reality, all sources are of finite size and thus cannot generate plane waves. However, under certain reasonable approximations, these electromagnetic waves can be approximated as plane waves. For example, if we take a spherical surface with oscillating currents, it generates what are known as spherical waves (since the wavefront of this wave is also a sphere). But if we look at a point far away from the sphere, in a small enough region of space, the spherical wave essentially looks like a plane wave. This is also why the sun's rays on this planet can be approximated by plane waves.

Questions

1. But what about a charged sphere whose radius is periodically oscillating about its fixed center?
2. What kind of electromagnetic waves does it radiate?

1.5 Electromagnetic Waves in Matter

In the previous section, we derived the expression for electromagnetic waves in vacuum. In this section, we will discuss the properties of these waves in matter, specially dielectrics. Any kind of matter is made up of atoms which in turn are made up of electrons, protons and neutrons. The protons and neutrons are tightly bound in the nucleus and do not respond to external electromagnetic fields unless the field strength is very high. For low and medium field strengths, usually its only the electrons that get affected and contribute to the electric/magnetic properties of matter. Certain types of materials respond strongly to electric fields, others to magnetic fields and still others to both electric and magnetic fields.

When an external electric/magnetic field is applied to a material, what happens is that the electron cloud of individual atoms gets distorted thereby leading to the emergence of an electric dipole moment or magnetic dipole moment or both (leading to non-zero bound charges and bound currents). This response is usually measured in terms of *polarization*. Electric polarization is the vector field that expresses the density of permanent or induced electric dipole moments in a dielectric material. It is a semi-classical phenomenon. Magnetic polarization is the vector field that expresses the density of permanent or induced magnetic dipole moments in a magnetic material. The origin of the magnetic moments responsible for magnetization can be either microscopic electric currents resulting from the motion of electrons in atoms, or the spin of the electrons or the nuclei. Net magnetization results from the response of a material to an external magnetic field, together with any unbalanced magnetic dipole moments that may be inherent in the material itself, for example, in ferromagnets. It is important to note that Bohr-van Leeuwen theorem shows that magnetism cannot occur in purely classical solids. Without quantum mechanics, there would be no diamagnetism, paramagnetism or ferromagnetism. Based on the polarization properties, matter can be broadly classified into the following types:

- **Dielectric:** Electric polarization is linearly proportional to the applied electric field (e. g., water and glass).
- **Paraelectric:** Electric polarization is a non-linear function of the applied electric field (e. g., $LiNbO_3$ above 1430 K).
- **Ferroelectric:** Spontaneous non-zero electric polarization even when the applied electric field is zero (e. g., Rochelle Salt (Potassium sodium tartrate)). Materials demonstrate ferroelectricity only below a certain phase transition temperature, called the Curie temperature, T_C, and are paraelectric above this temperature.

- **Diamagnetism:** Property of a material which causes it to create a magnetic field in opposition to an externally applied magnetic field ($\mu < 1$). These materials are repelled by magnetic fields.
- **Paramagnetism:** These materials have $\mu > 1$ and are attracted in the presence of externally applied magnetic fields. For diamagnetic and paramagnetic materials, the induced magnetic moment is linear in the field strength and rather weak. The attraction experienced by a ferromagnetic material is non-linear and much stronger, and hence, easily observed.
- **Ferromagnetism:** Basic mechanism by which certain materials form permanent magnets. The property of ferromagnetism is due to the direct influence of two effects from quantum mechanics: spin and Pauli exclusion principle.

The relation between the polarization and the bound charges/currents is given by

$$\vec{\nabla} \cdot \vec{P} = -\rho_b \qquad \text{Electric Polarization}$$

$$\frac{\partial \vec{P}}{\partial t} = \vec{J}_p \qquad \text{Polarization Curent}$$

$$\vec{\nabla} \cdot \vec{J}_p = -\frac{\partial \rho_b}{\partial t} \qquad \text{Continuity Equation}$$

$$\vec{\nabla} \times \vec{M} = \vec{J}_b \qquad \text{Magnetic Polarization}$$

$$\rho = \rho_f + \rho_b = \rho_f - \vec{\nabla} \cdot \vec{P}$$

$$\vec{J} = \vec{J}_f + \vec{J}_b + \vec{J}_p = \vec{J}_f + \vec{\nabla} \times \vec{M} + \frac{\partial \vec{P}}{\partial t} \tag{1.23}$$

where $\vec{P}, \vec{M}$ are the electric and magnetic polarization vectors respectively, ρ_f is the free charge, $\vec{J}_f$ the free current, ρ_b is the bound charge, $\vec{J}_p$ is the electric polarization current and $\vec{J}_b$ is the bound magnetic current. Note that free charges and free currents are usually present only in conductors. Using the above relations, Gauss' law can now be written as

$$\vec{\nabla} \cdot \vec{E} = \frac{\rho}{\epsilon_0} = \frac{1}{\epsilon_0}\left(\rho_f - \vec{\nabla} \cdot \vec{P}\right)$$

$$\Rightarrow \vec{\nabla} \cdot \vec{D} = \rho_f \tag{1.24}$$

where

$$\vec{D} = \epsilon_0 \vec{E} + \vec{P} \tag{1.25}$$

is known as the *electric displacement vector*. Similarly, Ampere-Maxwell law becomes

$$\vec{\nabla} \times \vec{B} = \mu_0 \left(\vec{J} + \vec{\nabla} \times \vec{M} + \frac{\partial \vec{P}}{\partial t}\right) + \mu_0 \epsilon_0 \frac{\partial \vec{E}}{\partial t}$$

$$\Rightarrow \vec{\nabla} \times \vec{H} = \vec{J}_f + \frac{\partial \vec{D}}{\partial t} = \vec{J}_f + \vec{J}_d \tag{1.26}$$

where $\vec{J}_d$ is the displacement current and

$$\vec{H} = \frac{\vec{B}}{\mu_0} - \vec{M} \tag{1.27}$$

which is also referred to as the magnetic field or the H-field in order to avoid confusion with the B-field. If the material under consideration is linear, we can write $\vec{P} = \epsilon_0 \chi_e \vec{E}$ and $\vec{M} = \chi_m \vec{H}$, we get

$$\begin{aligned} \vec{D} &= \epsilon \vec{E} \\ \vec{H} &= \frac{\vec{B}}{\mu} \end{aligned} \tag{1.28}$$

where χ_e, χ_m are the electric and magnetic susceptibilities, $\epsilon = \epsilon_0 (1 + \chi_e)$ is the permittivity and $\mu = \mu_0 (1 + \chi_m)$ is the permeability.

Among the four Maxwell's equations (Eq. 1.1), only two contain terms that depend on charge density and current density. Thus, for electromagnetic waves in matter only these two need to be written in terms of $\vec{D}$ and $\vec{H}$. The remaining two remain as it is. Thus, the Maxwell's equations in matter (assuming linearity) are

$$\begin{aligned} \vec{\nabla} \cdot \vec{D} &= \rho_f \\ \vec{\nabla} \times \vec{E} &= -\frac{\partial \vec{B}}{\partial t} \\ \vec{\nabla} \cdot \vec{B} &= 0 \\ \vec{\nabla} \times \vec{H} &= \vec{J}_f + \frac{\partial \vec{D}}{\partial t} \end{aligned} \tag{1.29}$$

It is important to note that the Maxwell's equations in matter (Eq. (1.29)) are exactly same as the general Maxwell's equations (Eq. (1.1)). But the reason for writing it in the above form is that the total charge density and total current density in a certain medium is usually unknown apriori. What is known is only the free charge density and free current density. Hence, writing it in the above form makes calculations relatively easier. After calculating $\vec{D}$ and $\vec{H}$ using the above equations, one can certainly go ahead and calculate the total charge and current density. However, this procedure works only if the medium is linear, i. e., the polarization depends linearly on the electric field. As mentioned earlier, there are many materials that show nonlinear behavior if the electric field strength is high enough. The electromagnetic field propagation in such media cannot be handled by the above equations and one needs to use more advanced tools.

Using Eq. (1.29), we will now derive the expression for an electromagnetic wave in dielectrics. For these materials, the free charge and free current density is zero. These two quantities are non-zero for conductors and will be dealt with in detail in Chapter 2.

Putting $\rho_f = 0 = \vec{J}_f$ in Eq. (1.29) and writing $\vec{D} = \epsilon\vec{E}$ and $\vec{H} = \vec{B}/\mu$, we get

$$\begin{aligned}\vec{\nabla}\cdot\vec{E} &= 0\\ \vec{\nabla}\times\vec{E} &= -\frac{\partial\vec{B}}{\partial t}\\ \vec{\nabla}\cdot\vec{B} &= 0\\ \vec{\nabla}\times\vec{B} &= \mu\epsilon\frac{\partial\vec{E}}{\partial t}\end{aligned} \tag{1.30}$$

where we have assumed ϵ and μ to be constants. Taking curl of the second equation above and substituting the fourth equation, we get

$$\begin{aligned}\vec{\nabla}\times\left(\vec{\nabla}\times\vec{E}\right) &= -\vec{\nabla}\times\frac{\partial\vec{B}}{\partial t}\\ \Rightarrow \vec{\nabla}\left(\vec{\nabla}\cdot\vec{E}\right) - \nabla^2\vec{E} &= -\mu\epsilon\frac{\partial^2\vec{E}}{\partial t^2}\\ \Rightarrow \nabla^2\vec{E} - \mu\epsilon\frac{\partial^2\vec{E}}{\partial t^2} &= 0\end{aligned} \tag{1.31}$$

which is the same equation as Eq. (1.17) except for the presence of $\mu\epsilon$ instead of $\mu_0\epsilon_0$. Hence, the speed of propagation of electromagnetic waves in dielectrics is

$$v = \frac{1}{\sqrt{\mu\epsilon}} = \frac{\omega}{k} = \frac{c}{n} \tag{1.32}$$

instead of $c = 1/\sqrt{\mu_0\epsilon_0}$ which is the speed of light in vacuum. Here, n is the refractive index of the dielectric medium. For most dielectrics, $\epsilon > \epsilon_0$ and $\mu \approx \mu_0$, which implies $v < c$. Hence, in most dielectrics, electromagnetic waves travel at a speed slower than than their speed in vacuum.

Questions

1. Do electromagnetic waves truly travel slower in dielectric media?
2. Or is it just an illusion?

Hint: Read chapter 31 of The Feynman Lectures on Physics, Volume 1.

The expressions for the electric and magnetic fields in this case can be derived by using the exact same procedure as was used to derive Eqs. (1.20) and (1.21). Here also, the direction of propagation, electric field and magnetic field are all mutually perpendicular to each other. It is important to note that in this section, it has been assumed that the dielectric is isotropic and homogeneous due to which its permittivity is just a constant scalar. However, for an anisotropic and/or inhomogeneous medium, the permittivity is a tensor and can depend on both the spatial and temporal coordinates. In that case, the relation $\vec{D} = \epsilon\vec{E}$ is valid only in the frequency domain and in the time/space-domain, the displacement vector is given by a convolution of the permittivity and electric field vector. As it turns out, the frequency dependence of ϵ is crucial for plasmonics and will be discussed in detail in Chapter 2.

In the previous section, we have represented the Maxwell's equations taking into account the electric and magnetic properties of linear media and then derived the equations for the electromagnetic waves in these materials. Using these equations, we can completely determine the nature of electromagnetic waves if the entire space consists of only one kind of material. However, this is seldom the case in practice. Specifically, plasmonics is the study of electromagnetic waves which propagate on the interface between a conductor and a dielectric. Hence, it is important to understand the process of finding the expressions for electromagnetic fields when the space is occupied by materials of two or more varieties. Within each of these materials, the electric and magnetic field expressions are still completely determined by solutions of the Maxwell's equations as described in the previous sections. What is required is to find a way to ensure that these expressions agree at the interface between two different materials. This is described by what are known as the *boundary conditions* and can be derived using the Maxwell's equations (Eq. (1.29)). These conditions are given by

$$D_1^{\perp} - D_2^{\perp} = \sigma_f$$

$$B_1^{\perp} - B_2^{\perp} = 0$$

$$\vec{E}_1^{\parallel} - \vec{E}_2^{\parallel} = 0$$

$$\vec{H}_1^{\parallel} - \vec{H}_2^{\parallel} = \vec{K}_f \times \vec{n} \tag{1.33}$$

where σ_f is the free surface charge density (should not be confused with conductivity, σ), $\vec{K}_f$ is the free surface current density (should not be confused with the propagation constant, k), $\perp$ represents the direction perpendicular to the interface and $\parallel$ represents the direction parallel to the interface, $\vec{n}$ is the unit vector normal to the interface.

1.6 Snell's Law and Evanescent Waves

In the previous sections, we have discussed the propagation of plane electromagnetic waves in free space and dielectric medium. At the end of the previous section, we also saw how to match the electric and magnetic fields at the boundary between two media. A question that arises in this context is regarding the direction of propagation of the electromagnetic wave. When an electromagnetic wave in incident on one medium from another medium, does it keep traveling in the same direction or does the direction of propagation undergo a change? Also, is all the incident energy transmitted into the other medium or is a part of it reflected back? This situation is depicted in Figure 1.1 where the interface between the two media is taken to be perpendicular to the z-axis. Here, θ_i, θ_t and θ_r represent the incident angle, transmitted (or refracted) angle and reflected angle, respectively. With respect to this figure, the two questions we are asking

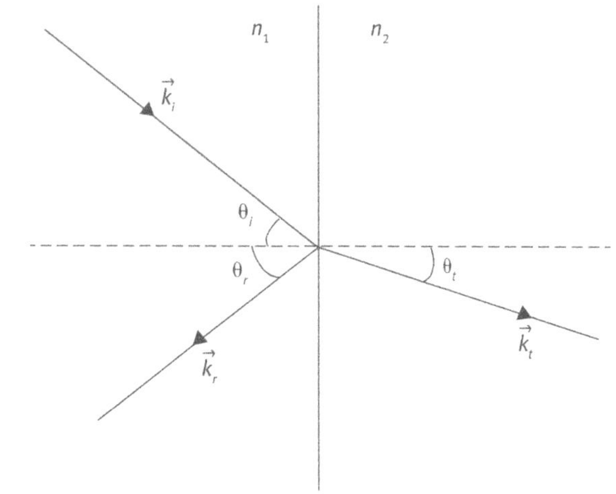

Fig. 1.1: This figure depicts an electromagnetic wave incident on the interface between two media making an angle of θ_i with the normal. A part of the incident energy is reflected back into the same medium at an angle $\theta_r = \theta_i$ with the normal and a part is transmitted into the second medium at an angle of θ_t with the normal. The incident and reflected waves have the same wave number, $\left|\vec{k}_i\right| = \left|\vec{k}_r\right|$, which is different from the wavenumber of the transmitted wave, $\left|\vec{k}_t\right|$. The relation between θ_t and θ_i is given by Snell's law, Eq. (1.35).

are:
1. What is the relation between θ_i, θ_t and θ_r?
2. Do the field strengths E_t and E_r become zero under certain situations?

There are several ways of deriving the relation between θ_i, θ_t and θ_r and we will here use the one based on translational symmetry arguments. The propagation vector $\vec{k}$ of the electromagnetic wave is associated with its momentum (photons!). Now the interface between the media cannot change the component of photon momentum in a direction parallel to the interface. Hence, by preservation of the parallel component of photon momentum, we get

$$k_1 \sin\theta_i = k_2 \sin\theta_t = k_1 \sin\theta_r \tag{1.34}$$

which implies $\theta_r = \theta_i$. Using Eq. (1.32), we can write $k_1 = \omega n_1/c$ and $k_2 = \omega n_2/c$, where n_1 and n_2 are the refractive indices of the two media. The frequencies of all the three waves (incident, transmitted and reflected) must be the same since the boundary conditions must be satisfied at all times. Thus, Eq. (1.34) becomes

$$n_1 \sin\theta_i = n_2 \sin\theta_t \tag{1.35}$$

which is known as the *Snell's law*.

In the context of the second question mentioned above, there are two possibilities to be considered. One is when there is no transmitted electromagnetic wave and the second is when there is no reflected electromagnetic wave. Without going into too much details, we will only mention the final result here and the readers are encouraged to do the detailed derivation themselves. It can be shown that when the transmitted wave direction and the reflected wave direction form a right angle between themselves, all the incident energy is transmitted into the other medium without any reflection. When this condition is met, we get $\theta_t = \pi/2 - \theta_r = \pi/2 - \theta_i$. The incident angle for which this happens is called the *Brewster's angle* and is usually denoted by θ_B. Substituting this in the Snell's law (Eq. (1.35)), we get

$$\begin{aligned} n_1 \sin\theta_B &= n_2 \sin\left(\frac{\pi}{2} - \theta_B\right) \\ \Rightarrow \theta_B &= \tan^{-1}\frac{n_2}{n_1} \end{aligned} \tag{1.36}$$

Thus, when light is incident from air ($n_1 \approx 1$) to glass ($n_2 \approx 1.5$), the Brewster's angle is approximately $5.36\,^\circ$. And for the reverse case (light incident from glass to air), θ_B is approximately $33.7\,^\circ$.

Now let us take the other case, i. e., when there is no transmitted electromagnetic wave. This condition is much easier to understand. From Eq. (1.35), we can write

$$\theta_t = \sin^{-1}\left(\frac{n_1}{n_2}\sin\theta_i\right) \tag{1.37}$$

We know that the sine function can only take values in the range $[-1, 1]$. Hence the inverse of sine function is valid only when the argument is in this range. This implies that a valid θ_t exists only when $(n_1/n_2)\sin\theta_i \leq 1$, assuming $\theta_i \in [0, \pi/2]$. The angle beyond which the transmitted wave ceases to exist is known as a the *critical angle* and is given by

$$\theta_c = \sin^{-1}\frac{n_2}{n_1} \tag{1.38}$$

This phenomenon is known as *total internal reflection* (TIR) and is of immense importance in optical fibers. It is important to note that a valid θ_c exists only when $n_1 > n_2$, i. e., when the light is incident from a rarer medium to a denser medium. On the other hand, there is no such restriction for the Brewster's angle to exist. For light propagating from glass to air, this critical angle is approximately $41.8\,^\circ$.

Let us look at this phenomenon of total internal reflection a little more closely. It turns out that if we write the expressions for the electric and magnetic fields, it becomes impossible to satisfy the boundary conditions (Eq. (1.33)) unless there is a nonzero field present in the second medium with refractive index n_2. This is quite strange since we just saw above that if the incident angle is above the critical angle, there is no transmitted electromagnetic wave in the second medium! This brings us to a very interesting and important concept known as *evanescent waves*. When $\theta_i > \theta_c$, although there is no electromagnetic energy transmitted into the second medium, there is still an electromagnetic wave that exists in this medium and propagates along the interface between the two media. An evanescent wave is a *near-field* wave with an intensity that exhibits *exponential decay without absorption* as a function of the distance from the boundary at which the wave was formed. In order to understand this further, let us go back to Eq. (1.37). As noted above, we get a valid value for θ_t only when the argument of the $\sin^{-1}$ is in the range $[-1, 1]$. However, this is true only when we insist on θ_t being a real valued quantity. If we relax this criteria and allow θ_t to be complex, we get a valid value for all values of the argument. Thus, we get

$$\begin{aligned}\sin\theta_t &= \frac{n_1}{n_2}\sin\theta_i > 1\\ \cos\theta_t &= \sqrt{1-\left(\frac{n_1}{n_2}\sin\theta_i\right)^2} \in \mathbb{C}\end{aligned} \tag{1.39}$$

where $\mathbb{C}$ denotes the set of all complex numbers. Let us now write the wave vector in the second medium as a vector sum of its components along the x- and z-axis,

$$\begin{aligned}\vec{k}_2 &= k_{2,x}\hat{x} + k_{2,z}\hat{z}\\ &= k_2\sin\theta_t\hat{x} + k_2\cos\theta_t\hat{z}\end{aligned} \tag{1.40}$$

Using Eq. (1.39), we get

$$k_{2,x} = k_2\sin\theta_t > k_2$$

$$\begin{aligned}k_{2,z} &= \sqrt{k_T^2 - k_x^2}\\ &= \frac{j\omega}{c}\sqrt{n_1^2\sin^2\theta_i - n_2^2}\end{aligned} \tag{1.41}$$

where $j = \sqrt{-1}$. Thus, $k_{2,x}$ is purely real and $k_{2,z}$ is purely imaginary. This implies that the evanescent wave in the second medium propagates along the interface in the x-direction and is exponentially decaying in the z-direction. Evanescent waves have immense importance in various applications, namely near-field antennas, waveguide filters, wireless energy transfer, detecting eavesdropping in optical fibers and also in plasmonics which we will discuss in later chapters.

Here it is important to note that the evanescent wave is decaying along the perpendicular only in the rarer medium and the electromagnetic wave continues to propagate without decay in the denser medium from where the wave was incident. A similar phenomenon of decay only one side called the *skin effect* is also found in electromagnetic waves incident from dielectrics on to metals. However, in this case the electromagnetic energy is lost/dissipated in the metal unlike the case of evanescent waves which is lossless. Both these effects are different from a surface plasmon wave which is lossless but decays on both sides of the interface between a dielectric and metal. The same metal-dielectric interface can thus hold both these waves: lossy skin effect and lossless plasmon wave. Which of these gets manifested depends on the frequency of the electromagnetic wave. The lossy skin effect happens at frequencies lower than terahertz range. And the plasmon wave exists at much higher optical and ultraviolet frequencies. In between these two limits, most metals behave like perfect electric conductors and simply reflect almost all of the electromagnetic energy incident on them. We will discuss further details in later chapters.

1.7 Group and Phase Velocity

As described in the previous section, the speed of electromagnetic waves in dielectrics $\left(v = \omega/k = 1/\sqrt{\epsilon\mu}\right)$ is usually lower than that in free space $\left(c = 1/\sqrt{\epsilon_0\mu_0}\right)$. This is because the relative permittivity $\left(\epsilon/\epsilon_0\right)$ of dielectrics is usually greater than one and its relative permeability is almost one. However, there are certain materials in which the speed of electromagnetic waves can be greater than the speed of light in vacuum since the relative permittivity in these materials can be smaller than one. This happens in metals in the frequency range of interest in plasmonics and poses a fundamental challenge since according to Einstein's special relativity, nothing in the universe can travel faster than $c = 3 \times 10^8 \mathrm{m/s}$. This brings us to an important concept known as *group velocity* which is different from the *phase velocity* $\left(v = \omega/k\right)$ described above.

It is observed in nature that the materials which have a phase velocity greater than light do so only in a certain frequency range. And in this frequency range, the permittivity of the material is found to be heavily dependent on frequency. Thus, if multiple electromagnetic waves of different frequencies (in this range) are traveling in such a material, they will all have different phase velocities. Now, if these frequencies are widely separated, the corresponding velocities are also widely separated and thus it is unlikely that these bunch of waves carry any meaningful information through this material. For there to be a meaningful transfer of information/energy through the material in a recoverable manner, the spread of frequencies should be within certain limits. If this condition is met, it can be shown that the bunch of waves (each having different phase velocities) moves through the material collectively at what is known as the *group velocity*. This concept is fairly general and applies to all kinds of wave phenomenon.

Consider a collection of waves moving along the x-axis and given by the expression

$$u(x, t) = \frac{1}{\sqrt{2\pi}} \int_{-\infty}^{\infty} A(k)\, e^{ikx - i\omega(k)t} dk \tag{1.42}$$

where $A(k)$ denotes the amplitude of individual wave components and ω/k is a function of k. If the spread of frequencies is small, then $A(k)$ is sharply peaked around some value $k = k_0$ and then we can do a Taylor expansion of ω around that point to get

$$\omega(k) = \omega(k_0) + \left.\frac{d\omega}{dk}\right|_{k_0} (k - k_0) + \cdots \tag{1.43}$$

Substituting Eq. (1.43) in Eq. (1.42), we get

$$\begin{aligned} u(x,t) &\simeq \frac{1}{\sqrt{2\pi}} \int_{-\infty}^{\infty} A(k)\, e^{ikx - i\omega_0 t - i\omega_0' kt + i\omega_0' k_0 t} dk \\ &= \frac{e^{i(\omega_0' k_0 - \omega_0)t}}{\sqrt{2\pi}} \int_{-\infty}^{\infty} A(k)\, e^{i(x - \omega_0' t)k} dk \\ &= u\left(x - \omega_0' t, 0\right) e^{i(\omega_0' k_0 - \omega_0)t} \end{aligned} \tag{1.44}$$

where $\omega_0 = \omega(k_0)$ and $\omega_0' = (d\omega/dk)\Big|_{k_0}$. Thus, we can see from Eq. (1.44) that the wave packet is essentially traveling along the x-axis with a velocity given by

$$v_g = \frac{d\omega}{dk}\Big|_{k_0} \tag{1.45}$$

which is known as the *group velocity*. And in the frequency range for which the phase velocity is larger than c in certain materials, it turns out that the group velocity is smaller than c. And this is what prevents Einstein's theory from being violated since it is the speed of information transfer that is of actual interest rather than the speed of phase transfer. The phase of plane wave carries no information and hence can travel at any speed without violating Einstein's theory. It must be noted that this is a heuristic argument and needs to be settled by something more rigorous.

The above description of group velocity is valid for one-dimensional wave propagation. But for propagation in two or three dimensions, we also need to specify the direction of the group velocity as it is not necessary that this direction will be same as that of phase velocity, which is always along $\vec{k}$. For a 2D propagation, the group velocity is given by

$$\vec{v}_g = \frac{\partial \omega}{\partial k_x}\hat{x} + \frac{\partial \omega}{\partial k_y}\hat{y} \tag{1.46}$$

and a similar expression can be written for 3D propagation. Let us now try to graphically understand the direction of group velocity. For this we need to draw what are known as *equifrequency* (or *isofrequency*) contours as shown in Figure 1.2. The lines in this figure connect the points in (k_x, k_y) space which correspond to the same frequency, ω. If we now pick up any one of these lines corresponding to a certain ω_0, the equation for this line becomes $\omega(k_x, k_y) = \omega_0$. If we now take a total derivative of this equation with respect to k_x, we get

$$\begin{aligned} &\frac{\partial \omega}{\partial k_x} + \frac{\partial \omega}{\partial k_y}\frac{dk_y}{dk_x} = 0 \\ \Rightarrow &\left(\frac{dk_y}{dk_x}\right)\left(\frac{\partial \omega/\partial k_y}{\partial \omega/\partial k_x}\right) = -1 \end{aligned} \tag{1.47}$$

In this equation, dk_y/dk_x is the slope of the tangent to the equifrequency contours, and from Eq. (1.46) we can see that $\left(\frac{\partial\omega/\partial k_y}{\partial\omega/\partial k_x}\right)$ is the slope of the line along the direction

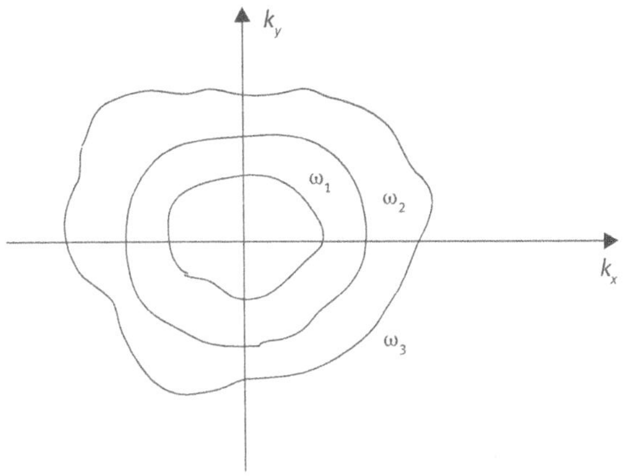

Fig. 1.2: This figure shows the equifrequency contours for the dispersion relation of an electromagnetic wave. At any point on this $k_x - k_y$ plane, the group velocity of the wave is along the direction of normal to the contour passing through that point.

of the group velocity. Thus, Eq. (1.47) clearly tells us that the direction of the group velocity is normal to the tangent of the equifrequency contours. So, the only situation where the group velocity and phase velocity will be in the same direction is when the equifrequency contours are given by concentric circles centered on the origin. In all other cases, the direction of the group velocity is likely to be different from that of the phase velocity. This is a very intriguing concept since this implies that the energy does not travel in the same direction as the phase! As they say, fact is stranger than fiction!

1.8 Waveguides

Another very important concept in electromagnetics which is also related to plasmonics, as we will see later, is that of waveguides. So far we have discussed electromagnetic waves freely traveling through vacuum and various kinds of matter. But in certain applications, we may want the EM wave to travel along a specific direction so as to minimize loss of energy and also prevent health hazards. Some applications require very high amount of EM energy and if these are allowed to propagate freely, they can cause a lot of damage to people and other instruments. This directionality to EM waves is provided through waveguides and optical fibers. These two work on very different principles. Waveguides work on the principle of total reflection from metallic surfaces, while optical fibers work on the principle of total internal reflection in dielectrics.

As discussed in Section 1.6, when an EM wave is incident from a denser to a rarer medium, it undergoes total internal reflection when the angle of incidence is larger than the critical angle. Optical fibers make use of this phenomenon by having a core of high refractive index surrounded by a cladding of lower refractive index, as shown in the Figure 1.3. And light is made to incident in the core region such that its angle of incidence on the cladding is higher than the critical angle. It is these optical fibers that are used for internet connectivity and are very efficient in transmitting data over hundreds of kilometers at a very fast rate as required for modern communication systems. However, optical fibers are suitable for transmission of only low power and high frequency (optical range) EM waves. For many applications we need to work at the GHz range and also at high power, where optical fibers are very inefficient. In such cases, we need to use another electromagnetic device known as waveguides.

An electromagnetic waveguide is essentially a tubular structure bounded by metallic conductors on four sides allowing for passage of EM waves through two sides. Figure 1.4 shows two typical waveguides with rectangular and circular cross-section. A waveguide can be hollow or also contain a central conductor (e. g., coaxial cables). And the cross-section of a waveguide can in general be quite complex. Here we will discuss only hollow waveguides with rectangular cross-section. It is important to note that metals behave as good reflectors of EM waves only in a certain frequency range of the order of GHz and THz (as will be discussed in the next chapter). So although our analysis in this section is theoretically valid for all frequencies, it is practically relevant only for this particular frequency range.

The electric and magnetic fields inside the waveguide must satisfy the equation for an EM wave given by Eq. (1.31),

$$\vec{\nabla}^2 \vec{E} - \mu\epsilon \frac{\partial^2 \vec{E}}{\partial t^2} = 0.$$

Taking a Fourier transform in the time-domain, we get

$$\vec{\nabla}^2 \vec{E} + \omega^2 \mu\epsilon \vec{E} = 0.$$

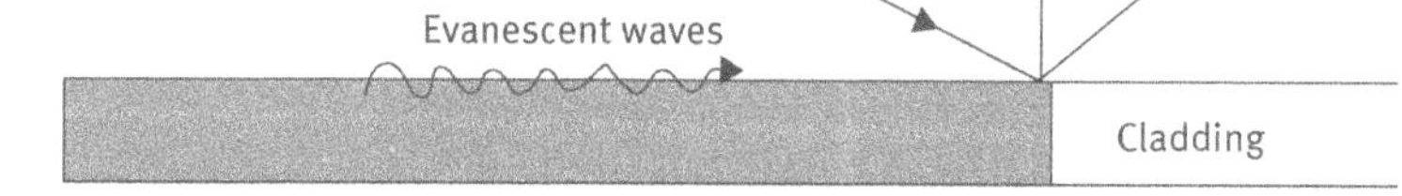

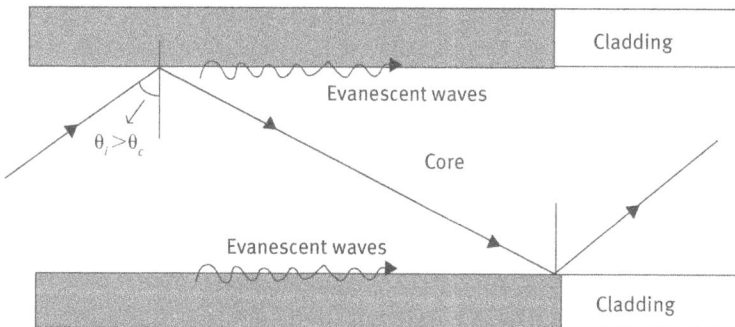

Fig. 1.3: This figure is an illustration of an optical fiber. The core has a higher refractive index than the cladding due to which we have the phenomenon of total internal reflection. However, some of the electromagnetic energy also travels in the form of evanescent waves in the cladding.

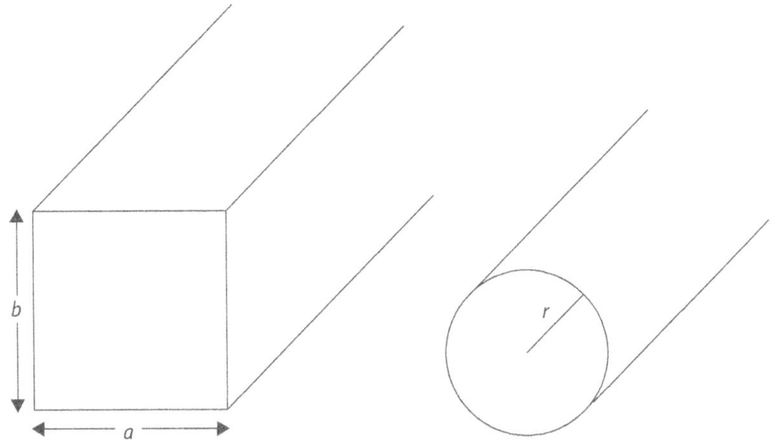

Fig. 1.4: This figure shows two simple examples of waveguides made of conducting material.

For further analysis, let us consider the component of the electric field along the direction of propagation of the wave, $\hat{z}$. Once we are able to solve for this component, the other components of the electric and magnetic field can be easily obtained using the Maxwell's equations. The equation for the z-component of the electric field is given by

$$\frac{\partial^2 E_z}{\partial x^2} + \frac{\partial^2 E_z}{\partial y^2} + \frac{\partial^2 E_z}{\partial z^2} + \omega^2 \mu \epsilon E_z = 0 \tag{1.48}$$

This equation can be solved using the method of *separation of variables*, in which we assume that the function we are trying to solve for can be written as the product of three functions each of which is a function of only one of the variables. In our case, this leads to the assumption that E_z can be written as

$$E_z(x, y, z) = X(x)\, Y(y)\, Z(z) \tag{1.49}$$

Substituting this in Eq. (1.48), we get

$$\frac{X''}{X} + \frac{Y''}{Y} + \frac{Z''}{Z} = -\omega^2 \epsilon \mu \tag{1.50}$$

where the prime represents derivative with respect to the variable of which the corresponding quantity is a function. Now each term on the LHS of Eq. (1.50) is a function of an independent and different variable. If their sum is a constant, then each term must also be a constant. Thus, we can write

$$\frac{X''}{X} = -k_x^2 \qquad \frac{Y''}{Y} = -k_y^2 \qquad \frac{Z''}{Z} = -k_z^2 \tag{1.51}$$

where $k_x^2 + k_y^2 + k_z^2 = \omega^2 \epsilon \mu$. In these waveguides, k_x, k_y are always real but k_z can be imaginary depending on whether the wave is propagating or evanescent. The most general solution for Eq. (1.51) is given by

$$E_z(x, y, z, t) = \sum_{k_x, k_y} [A\cos(k_x x) + B\sin(k_x x)] \times [C\cos(k_y y) + D\sin(k_y y)]\, e^{i(k_z z - \omega t)} \tag{1.52}$$

where the coefficients A, B, C, D also depend on k_x, k_y. In order to solve for these coefficients, we need to use the boundary conditions given by Eq. (1.33). According to these conditions, the parallel component of the electric field must be continuous across the boundary. Now since our metal is a perfect electric conductor at our frequency of interest, the electric field inside it must be zero. And hence, the parallel component of the electric field must also go to zero at the boundaries. And as it turns out, E_z is parallel to all the four boundaries! Thus, in order to ensure that E_z satisfies the boundary conditions at $x = 0$, $x = a$, $y = 0$ and $y = b$, we must have $A = 0 = C$ and $\sin(k_x a) = 0 = \sin(k_y b)$. This implies that k_x and k_y must be integer multiples of π/a and π/b respectively,

$$k_x = \frac{m\pi}{a} \qquad k_y = \frac{p\pi}{b}$$

where $m, p \in Z$. So, k_z is given by

$$k_z = \sqrt{\omega^2\mu\epsilon - \left(\frac{m\pi}{a}\right)^2 - \left(\frac{p\pi}{b}\right)^2} \tag{1.53}$$

As can be seen from this equation, for a given value of ω, k_z is real only for those values of m, p for which the term under the square root is positive. The values of m, p for which k_z is imaginary constitute the evanescent waves. Thus, for any frequency, the EM wave inside a waveguide is in general a superposition of propagating and evanescent waves. However, for certain frequencies there may be no m, p for which there is a real k_z. Since $A = 0 = C$ in Eq. (1.52), the values for m, p must be larger than zero since otherwise E_z will become zero. Hence, the lowest frequency for which this EM wave (with non-zero E_z) can exist in the waveguide is given by

$$\omega_{c1} = \sqrt{\frac{1}{\mu\epsilon}\left[\left(\frac{\pi}{a}\right)^2 + \left(\frac{\pi}{b}\right)^2\right]} \tag{1.54}$$

If we allow the E_z to be zero, then H_z must be non-zero. It is not possible for both E_z and H_z to be zero in a waveguide unless it has a central conductor (e. g., coaxial cable). In our hollow waveguide, if we choose the option of allowing E_z to go to zero but keep a non-zero H_z, a similar analysis as above can be carried out and it leads to a lower cut-off frequency

$$\omega_{c2} = \frac{\pi}{a}\sqrt{\frac{1}{\mu\epsilon}} \tag{1.55}$$

where we have assumed $a > b$ (without loss of generality). For the case of EM waves in a waveguide which has non-zero E_z and zero H_z, the magnetic field is perpendicular to the direction of propagation of the wave. Hence, these modes are known as *transverse magnetic* (TM). Similarly, the case of EM waves which has non-zero H_z and zero E_z is known as *transverse electric* (TE). The EM wave component corresponding to a particular value of m, p is known as the TE_{mp} or TM_{mp} mode as the case may be. So, in a hollow rectangular waveguide, the mode with the lowest cutoff frequency is the TE_{10} mode and the corresponding frequency is given by Eq. (1.55). An EM wave with frequency below this cut-off cannot propagate inside the waveguide and forms evanescent waves instead. The decay length of these evanescent waves are again given by Eq. (1.53) with the only difference that k_z will now be imaginary. These evanescent waves are very important in the propagation of surface plasmons at GHz frequencies as will be discussed in Chapter 5.

2 Electromagnetic Properties of Metals

In the previous chapter, we have introduced Maxwell's equations and studied the behavior of electromagnetic waves in free space and dielectrics. We have also introduced the concept of permittivity and permeability. In this chapter, we will take this further and primarily study two concepts: origin of permittivity and electromagnetic waves in conductors (metals). By origin of permittivity, we mean the dependence of permittivity on the fundamental parameters of the material (dielectric or conductor). We will derive the expression for permittivity starting from Newton's force equation and try to understand its behavior in various frequency ranges. The nature of electromagnetic waves in metals is very different from that in dielectrics due to the presence of free electrons. These free electrons lead to a non-zero conductivity which manifests as the imaginary part of the effective permittivity which is now complex (unlike the dielectric case where it is real for most frequencies). Due to this, the electromagnetic waves in conductors are spatially damped and lead to loss of electromagnetic energy. As we will see, there is also a frequency range in which the permittivity of conductors becomes approximately real and negative. It is this frequency range (usually around the optical or ultraviolet region) that is of primary interest in the context of plasmonics.

2.1 Origin of Permittivity

In a metal, there are two kinds of electrons: bound and free. Bound electrons are the ones that are tightly bound to the nucleus of the atom and move within the atomic orbitals. Free electrons are the ones present in the conduction band of metals and responsible for their electrical and thermal conductivity. In order to find the permittivity of any material, we need to first write an equation governing the motion of these electrons. Strictly speaking, the dynamics of both these kinds of electrons is governed by quantum mechanics. However, that is quite complicated in general and very difficult to solve analytically. For most practical purposes, it is sufficient to use what is known as the semi-classical approach. We use the classical Newton's equation of motion for electron dynamics, but the parameters of this equation need to calculated quantum mechanically. Also, we will only see how to solve the equations for a given set of parameters without going into their quantum mechanical foundations.

In the semi-classical approach, the motion for an electron can be considered to be equivalent to a damped spring in an external force field as shown in the Figure 2.1. Neglecting the nonlinear contributions of the spring force, the 1D equation of motion of an electron bound by a damped simple harmonic force and acted upon by an oscillatory electric field $E(t)$ is

$$\ddot{x} + \gamma\dot{x} + \omega_0^2 x = -\frac{e}{m}E(t) \tag{2.1}$$

https://doi.org/10.1515/9783110570038-035

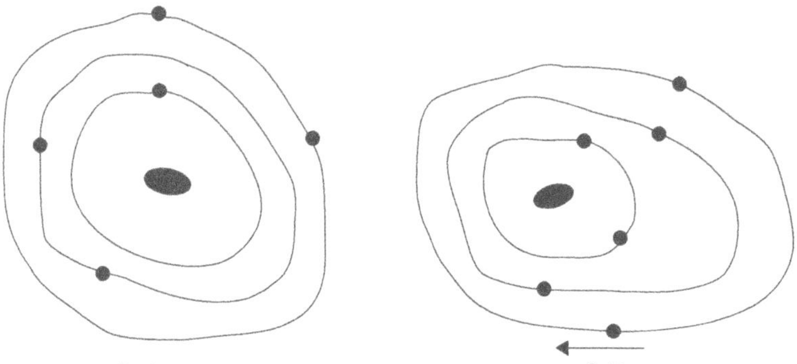

Fig. 2.1: This figure shows the polarisation of an atom due to the presence of electric field.

where γ is the damping constant, ω_0 is the binding frequency or natural frequency of oscillation of the electron, $-e$ is the electron charge and m is the electron mass. Note that the parameters γ and ω_0 are to be estimated using quantum mechanics. Also, in the above, magnetic force has been neglected since we assume the electron velocity to be much smaller than speed of light. Here it is important to note that x is not the position of the electron, but that of the center of charge of the electron cloud.

Taking a Fourier transform of Eq. (2.1) with respect to time, the dipole moment contributed by one electron is obtained to be

$$p(\omega) = -ex(\omega) = \frac{e^2}{m} \frac{E(\omega)}{\omega_0^2 - \omega^2 - i\omega\gamma} \tag{2.2}$$

An atom usually has many electrons and the total dipole moment is just a simple sum of the dipole moment due to each electron. If there are N molecules per unit volume with Z electrons per molecule, and instead of a single binding frequency for all, there are f_j electrons per molecule with binding frequency ω_j and damping constant γ_j such that $\sum_j f_j = Z$, then the total dipole moment per unit volume is given by

$$\begin{aligned} P(\omega) &= N \sum_j f_j p_j \\ &= \frac{e^2 N E(\omega)}{m} \sum_j \frac{f_j}{\omega_j^2 - \omega^2 - i\omega\gamma_j} \end{aligned} \tag{2.3}$$

Using Eq. (1.25), we can write the electric displacement to be

$$\begin{aligned} D &= \epsilon_0 E + P \\ &= E(\omega) \left(\epsilon_0 + \frac{Ne^2}{m} \sum_j \frac{f_j}{\omega_j^2 - \omega^2 - i\omega\gamma_j} \right) \end{aligned} \tag{2.4}$$

Now, the permittivity of the material is given by

$$\begin{aligned}\epsilon &= \frac{D}{E} \\ &= \epsilon_0 + \frac{Ne^2}{m}\sum_j \frac{f_j}{\omega_j^2 - \omega^2 - i\omega\gamma_j}\end{aligned} \tag{2.5}$$

which is complex even for dielectrics. However, the imaginary part of this permittivity is significant only for a very small frequency ranges close to the ω_j points. This is because γ_j is usually very small compared to ω_j. Thus, for the rest of the frequency range, the permittivity can be approximately taken to be a real quantity. Imaginary part of ϵ is large and positive only in the region of *anomalous dispersion* $\left(\partial\epsilon/\partial\omega < 0\right)$ and represents a dissipation of electromagnetic energy into the medium. This phenomenon is known as resonant absorption. If imaginary part of ϵ is negative, energy is given by the medium and amplification occurs as in a maser or laser. It can be clearly seen from Eq. (2.5) that for very large ω, $\epsilon/\epsilon_0 < 1$ but tends to 1 as $\omega \to \infty$. Thus, at very high frequencies of the electromagnetic wave, most materials behave like free space.

In the above derivation, one of the most important points to note is that the displacement vector, D, is a product of ϵ and E only in the frequency domain. In the time domain, D is thus given by a convolution of ϵ and E,

$$\begin{aligned}D(\omega) &= \epsilon(\omega)\,E(\omega) \\ \Rightarrow D(t) &= \epsilon(t) \star E(t)\end{aligned} \tag{2.6}$$

If $\epsilon(\omega)$ is approximately constant and independent of frequency, then $\epsilon(t)$ is an appropriately scaled Dirac delta function. In this case, the convolution in time-domain essentially becomes a multiplication of $E(t)$ with the constant scaling factor. For most dielectrics, $\epsilon(\omega)$ is approximately constant and we don't have to worry about this concept. But in the frequency range of interest in plasmonics, the permittivity depends heavily on frequency and here this concept becomes very important.

2.2 Permittivity and Conductivity of Conductors

In the previous section, we have derived the permittivity of dielectrics and seen how it depends on the parameters of the atoms that make up the dielectric material. The same expression can also be used to find the permittivity and conductivity of conductors. As mentioned earlier, a conductor contains two types of electrons: bound and free. For the free electrons, the binding frequency is zero ($\omega_0 = 0$). In this case, though the damping constant (γ_0) is non-zero, its origins are different from the damping constant of bound electrons. In the case of bound electrons, the damping constant is a purely quantum mechanical parameter and cannot be imagined in a classical context. For the case of free electrons, the damping constant arises out of collisions among the free electrons moving across the conductor.

The permittivity of conductors is thus given by writing Eq. (2.5) as a sum of two parts, one for the bound electrons and one for the free electrons,

$$\epsilon = \epsilon_b + i\frac{Ne^2 f_0}{m\omega(\gamma_0 - i\omega)} \tag{2.7}$$

where ϵ_b is the permittivity contribution of the bound electrons. Note that Eq. (2.7) represents the *effective permittivity* of conductors and it is ϵ_b which is usually called the permittivity. In order to extract the expression for conductivity from Eq. (2.7), let us go back to the Ampere-Maxwell equation

$$\begin{aligned}\vec{\nabla} \times \vec{H} &= \vec{J} + \frac{\partial \vec{D}}{\partial t} \\ &= \sigma\vec{E} - i\omega\epsilon_b\vec{E} \\ &= -i\omega\left(\epsilon_b + i\frac{\sigma}{\omega}\right)\vec{E}\end{aligned} \tag{2.8}$$

where we have used Ohm's law, $J = \sigma E$, and carried out Fourier transform in the time domain. Comparing the term in brackets in the RHS of Eq. (2.8) with the expression for effective permittivity given by Eq. (2.7), we can see that the expression for conductivity can be written as

$$\sigma = \frac{Ne^2 f_0}{m(\gamma_0 - i\omega)} \tag{2.9}$$

which is also known as the *Drude model* for free electrons. In reality, the problem of conductivity is a quantum mechanical one in which the Pauli principle plays a crucial role. The damping effects are a result of collisions that involve considerable momentum transfer. For copper, $N \approx 8 \times 10^{28}$ atoms/m^3 and at normal temperature $\sigma \approx 5.9 \times 10^7\ (\Omega\text{m})^{-1}$, which implies $\gamma_0 \approx 4 \times 10^{13}\ \text{s}^{-1}$ assuming $f_0 \sim 1$. Thus, up to the microwave region $\left(\omega \leq 10^{11}\ \text{s}^{-1}\right)$, conductivity of copper is essentially real and independent of frequency.

For frequencies much higher than γ_0, the effective of conductors becomes

$$\begin{aligned}\epsilon &= \epsilon_b - \frac{Ne^2 f_0}{m\omega^2} \\ &= \epsilon_b - \frac{\omega_p^2}{\omega^2}\epsilon_0\end{aligned} \tag{2.10}$$

where $\omega_p = \sqrt{Ne^2 f_0/(m\epsilon_0)}$ is the plasma frequency. For such high frequencies, $\epsilon_b \simeq \epsilon_0$ and the relative permittivity $\left(\epsilon_r = \epsilon/\omega_0\right)$ becomes

$$\epsilon_r = 1 - \frac{\omega_p^2}{\omega^2} \tag{2.11}$$

and it is this frequency range that is of interest in plasmonics. We will see in Chapters 3 and 4 that this is exactly same as the relative permittivity of plasmas, and that is where plasmonics gets its name from. For most metals, ω_p is in the optical and ultraviolet

Tab. 2.1: Table of plasma frequency and conductivity of common metals.

Metal	Plasma Frequency $(f_p = \omega_p/2\pi)$
Silver	2.3 PHz
Aluminium	3.7 PHz
Gold	2.1 PHz
Copper	1.9 PHz
Potassiium	0.9 PHz
Sodium	1.4 PHz
Platinum	1.2 PHz

Source: http://www.wave-scattering.com/drudefit.html

region (see Table 2.1) and the frequency of plasmon waves is slightly lower than ω_p since we require ϵ to be slightly negative. We will discuss details of these plasmon waves in Chapter 5. In the remaining portion of this chapter, we will discuss the nature of electromagnetic waves in a conductor with constant ϵ_b and σ.

2.3 Electromagnetic Waves in a Conductor

We can find an expression for the dispersion relation of electromagnetic waves in conductors by using the same method described in Section 1.5. The only difference is that in addition to permittivity, we now also have a non-zero conductivity that leads to very different properties of the electromagnetic waves. Taking the curl of Faraday's equation (Eq. (1.29)), substituting the Ampere-Maxwell equation and using the Poisson's equation in charge-free space $\left(\vec{\nabla} \cdot \vec{E} = 0\right)$, we get

$$\begin{aligned} \vec{\nabla} \times \left(\vec{\nabla} \times \vec{E}\right) &= -\vec{\nabla} \times \frac{\partial \vec{B}}{\partial t} \\ \Rightarrow \nabla^2 \vec{E} &= \mu\sigma \frac{\partial \vec{E}}{\partial t} + \mu\epsilon_b \frac{\partial^2 \vec{E}}{\partial t^2} \end{aligned} \tag{2.12}$$

Taking a Fourier transform of the above equation, we get

$$\tilde{k}^2 = \mu\epsilon_b\omega^2 + i\mu\sigma\omega \tag{2.13}$$

which implies that the wave number is no longer a real quantity. In order to separately analyze the real and imaginary parts of $\tilde{k}$, we now write

$$\tilde{k} = k + i\kappa \tag{2.14}$$

where k is the propagation constant and κ is the decay constant. This phenomenon is known as *skin effect* and leads to loss of electromagnetic energy inside the conductor. Here it is important to note an important difference from the case of evanescent waves described in Section 1.6. In the case of skin effect, the direction of propagation is same as the direction of decay. It is this property that leads to loss of electromagnetic energy in conductors. In the case of evanescent waves in dielectrics, the propagation direction is perpendicular to the direction of decay due to which there is no loss of electromagnetic energy.

In order to find separate expressions for the propagation constant and the decay constant, substituting Eq. (2.14) in Eq. (2.13), we get

$$k = \omega\sqrt{\frac{\epsilon_b\mu}{2}}\left(1+\sqrt{1+\left(\frac{\sigma}{\epsilon_b\omega}\right)^2}\right)^{1/2} \tag{2.15}$$

and

$$\kappa = \omega\sqrt{\frac{\epsilon_b\mu}{2}}\left(\sqrt{1+\left(\frac{\sigma}{\epsilon_b\omega}\right)^2}-1\right)^{1/2} \tag{2.16}$$

The inverse of the decay constant is known as the *skin depth*,

$$\delta = \frac{1}{\kappa} \tag{2.17}$$

and gives the scale length over which electromagnetic energy decays inside the conductor. Since the decay of the electric field strength is governed by $\exp\left(-z/\delta\right)$, we find that on traveling a distance of 3δ the field strength decays by about 95% of its original value. From Eq. (2.16) it is also clear that higher the value of σ, higher will be the value of κ and lower the value of δ. Thus, for a very good conductor ($\sigma \gg \epsilon_b\omega$), the skin depth is very small and reverse is the case for a poor conductor ($\sigma \ll \epsilon_b\omega$). In these two limiting conditions, Eqs. (2.15) and (2.16) can simplified to give:

- Good conductor ($\sigma \gg \epsilon_b\omega$):

$$\begin{aligned} k = 2\pi/\lambda &\approx \sqrt{\sigma\omega\mu/2} \\ \kappa &\approx \sqrt{\sigma\omega\mu/2} \approx k \\ \delta &\approx \sqrt{2/\sigma\omega\mu} \approx \lambda/2\pi \end{aligned} \tag{2.18}$$

- Poor conductor ($\sigma \ll \epsilon_b\omega$):

$$\begin{aligned} k &\approx \omega\sqrt{\epsilon_b\mu}\left(1+\frac{\sigma^2}{8\epsilon_b^2\omega^2}\right) \\ \kappa &\approx \sqrt{\mu\sigma^2/4\epsilon_b} \\ \delta &\approx \sqrt{4\epsilon_b/\mu\sigma^2} \end{aligned} \tag{2.19}$$

It is interesting to note that for the case of a good conductor, the skin depth is independent of the permittivity and for the case of a poor conductor, the skin depth is independent of frequency. Now since the definition of a good/poor conductor depends on frequency, we can change the above labels to ‘low frequency’ and ‘high frequency’. Thus, the same metal behaves as a poor conductor much above a certain frequency range and as a good conductor much below it. This boundary in frequency between these two kinds of behavior needs to be properly evaluated since, as described in the previous sections, the conductivity and permittivity themselves depend on frequency. What we usually call conductivity and permittivity of a metal (or any other substance for that matter) is its value for DC conditions ($\omega = 0$). At higher frequencies, these values obviously undergoes a change.

3 Plasma Kinetic Theory

3.1 Introduction

In the previous chapter, we have studied the dynamics of single particles in the presence of various kinds of electromagnetic (EM) fields. In this chapter, we will take this to the next level and study the dynamics of a collection of charged particles (also called a *plasma*) in these field configurations. For single particle motion, all that needs to be studied is the time-evolution of its position and velocity. However, when we have several particles, keeping track of the individual trajectories is very data intensive. So, we do a kind of averaging and for a given position and velocity, we study how many particles have positions and velocities close to these values. And this number keeps changing with time and essentially represents the density in phase space (configuration space is made of only position whereas phase space is made of position and velocity/momentum). This number is popularly called the *plasma distribution function* and usually denoted by $f(x, v, t)$ or $P(x, v, t)$. In this book, we will stick to f. In the language of probability theory, this quantity is known as the probability density function (PDF). Strictly speaking, f is a function of 7 variables, *viz.* three position variables, three velocity variables and one time variable. But in this chapter and next, we will only consider the 1D case for simplicity and hence we consider f to be a function of only 3 variables. For a given problem, we sometimes have several different possibilities for the function f and this choice is not always easy to make. However, in any case, the integral of f over x and v must be finite and constant for all time (unless we are studying an open system where number of particles can change with time).

Our function of interest, the plasma distribution function, depends on three independent parameters: position (x), velocity (v) and time (t). This is a significant difference from single particle motion since in that case, the position and velocity are dependent on time. But when we study collective motion, these two variables are as independent as time. Since we are interested in studying the distribution function for various kinds of electromagnetic fields, what we need now is an equation which can be solved to achieve this objective. Here, it is important to introduce the very important concept of *collisions*. Collisions essentially refer to the inter-particle interactions in the plasma. We will discuss this concept in some detail in later sections of this chapter, but for now it would be sufficient to note that plasmas are usually dilute systems and collisions can be neglected.

For the case of collisionless plasmas, the equation governing the plasma distribution function is luckily very easy to find and is known as the Liouville or Vlasov equation. Consider a certain number of particles occupying a region of phase space as shown in Figure 3.1. If we neglect inter-particle collisions, it can be shown that as this region moves with time, the total number of particles in this remains constant.

https://doi.org/10.1515/9783110570038-043

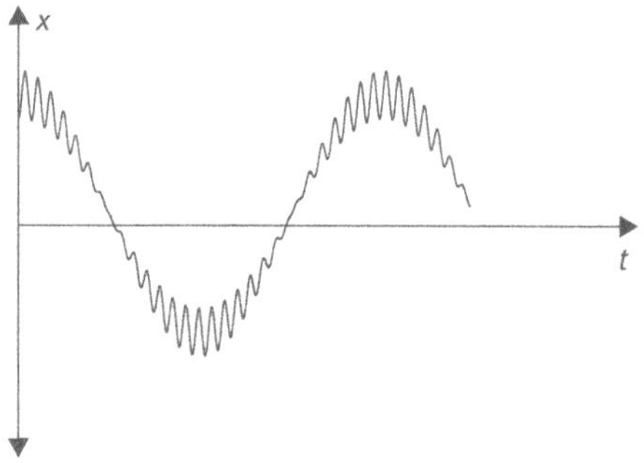

Fig. 3.1: This figure shows the path of a particle moving under the influence of time-periodic spatially non-uniform forces as given by Eq. (3.20).

This is known as Liouville's theorem. This essentially implies that

$$\frac{df}{dt} = 0 \qquad \text{(Liouville Equation)} \tag{3.1}$$

Using the chain rule, we can now write the above in terms of partial derivates to get

$$\frac{df}{dt} = \frac{\partial f}{\partial t} + \frac{dx}{dt}\frac{\partial f}{\partial x} + \frac{dv}{dt}\frac{\partial f}{\partial v}$$

$$\Rightarrow \frac{\partial f}{\partial t} + v\frac{\partial f}{\partial x} + a\frac{\partial f}{\partial v} = 0 \qquad \text{(Vlasov Equation)} \tag{3.2}$$

where $v = dx/dt$ is the velocity and $a = dv/dt$ is the acceleration usually given by the Lorentz force divided by mass (for the case of charged particle dynamics). It must be noted that although we have used $v = dx/dt$ in the derivation of Vlasov equation, we treat v and x as independent parameters while further analyzing this equation and its solutions. This is a subtle point and needs to be carefully thought about.

Now that we have an equation governing the dynamics of charged particles in electromagnetic fields, the next step would be to study the method of solving it. We will first discuss the general approach to solving this equation and then take up specific examples. As it turns out, there is no single way to solve the Vlasov equation for a general electromagnetic field. Also, for any given electromagnetic field, the Vlasov equation has infinitely many solutions (explained in the next section). The solution method adopted and the particular solution chosen varies widely according to the field. Some of these methods can be broadly classified into the following categories:

1. Exact solutions for time-independent electric fields
2. Linear response theory (plasma waves)
3. Ponderomotive theory

The above list is not exhaustive and only describes some of the basic methods of solving Vlasov equation. The entire theory is too vast and beyond the scope of this book.

We will only focus on the above categories to give the reader a flavor of the basic ideas and more advanced tools can be found in research papers. In this chapter, we have ignored collisions which form a very important component of plasma dynamics. We could ignore collisions since in many cases the plasma is dilute enough thereby making inter-particle interactions very rare. However, in several other cases, this may not be so making it necessary to consider the effect of collisions. These collisions are usually of two types: head-on short range collisions [1] and Coulombic long-range collisions. The head-on short range collisions have been thoroughly studied and are very well understood. However, the Coulombic long-range collisions are very difficult to handle mathematically and are an active area of research. Though strictly speaking, plasma collisions belong to the later category, several aspects of the mathematical modeling used for head-on collisions give good agreement with experimental findings. In this chapter, we have also ignored the magnetic field effects which are very important in many plasma phenomenon. However, these effects are not important in the context of plasmonics and the interested reader can see [2] for further details of these magnetic field effects.

Now we take each of these categories one by one and discuss the corresponding methods.

3.2 Exact Solutions for Time-Independent Electric Fields

If the plasma is in a time-independent electric field, the solution for the 1D Vlasov equation is quite simple. A careful look at Eq. (3.1) reveals that f is nothing but a *constant of motion*. As the name denotes, a constant of motion is a property of the particle which remains constant as the particle moves around in the given EM field. So, for a given EM field, if we can find a constant of motion, the plasma distribution function is any arbitrary function of this constant. That is why the Vlasov equation has infinitely many solutions for any given EM field. Now although this is a general idea and theoretically valid for any arbitrary EM field, it is extremely difficult to find such a constant for general time-dependent EM fields. That is why its use is usually limited to the case of time-independent electric fields and few types of time-periodic electric field as we will see later.

For the case of time-independent electric field, it is obvious that the particle energy (which is also the particle's Hamiltonian) is a constant of motion,

$$H = \frac{1}{2}mv^2 + qU(x) \tag{3.3}$$

where $U(x)$ is the electric potential, m the particle mass and q the charge. As mentioned above, any arbitrary function of H given by Eq. (3.3) is a valid solution of Eq. (3.2). Also, this H is not the only constant of motion and many others can be constructed. Now arises a very important question regarding the choice to be made.

Which of these infinite solutions is the correct representation of the plasma distribution? This dilemma cannot be resolved through the Vlasov equation and we must look beyond it. There are two approaches to resolve this: BGK approach and equilibrium solutions.

While solving Eq. (3.2), we assumed that our electric potential is already given to us. However, the plasma consists of charged particles which also have their own electric field. The electric field of all these particles adds up with the applied electric field and contributes to the total electric field. In the Vlasov picture, the choice of f determines what this induced field $[E(x)]$ must be and can be estimated using the Poisson's equation

$$\frac{\partial}{\partial x} E(x) = \frac{qn(x)}{\epsilon} \tag{3.4}$$

where ϵ is the permittivity and $n(x)$ is the plasma density obtained by integrating f over velocity. If the plasma has more than one species of charged particles, then the above equation needs to be suitably modified. Hence, the correct approach would be one in which an expression for U and f is obtained by solving the Vlasov and Poisson equation together. This method is described in a landmark paper published in 1957 and is popularly known as the BGK approach [3]. We will not go into this approach in this book since its not very relevant to the topic of plasmonics but readers are encouraged to go through this BGK paper in detail.

A much simpler solution is given by neglecting the induced field and considering only the Vlasov equation without bothering about the Poisson's equation. This is valid when the plasma is dilute enough and the applied field strength strong enough. Under these and few more conditions, it is known in statistical mechanics that the equilibrium distribution for time-independent systems is given by the Gibbs-Boltzmann distribution

$$f(x, v) = n_0 \sqrt{\frac{m}{2\pi\kappa T}} \exp\left(-\frac{H}{\kappa T}\right) \tag{3.5}$$

where H is the constant of motion given by Eq. (3.3), κ is the Boltzmann constant, T is the temperature and n_0 is the total number of particles in the system. There is an important catch here which is one of the yet unresolved problems in plasma physics. Strictly speaking, (3.5) holds when the particles experience head-on collisions like what happens in a gas of neutral particles. But in the case of the plasma, the interactions are long-range due to the Coulomb force. Few attempts have been made at arriving at the equilibrium distribution function for these long-range forces, but the resultant expressions are too complex to be of any practical use. However, in several plasma experiments, the particle distribution has been found to be close to a Gaussian. Hence, most researchers working in this field invariably choose the Gibbs-Boltzmann distribution for their work.

3.3 Linear Response Theory (Plasma Waves)

As described in the previous section, the plasma distribution function for a time-independent electric field is simply given by any function of its Hamiltonian given by Eq. (3.3) and for most practical purposes we choose the Gibbs-Boltzmann distribution given by Eq. (3.5). A question of immense importance in plasma physics is what happens when this solution is slightly perturbed. These perturbations (also called *plasma waves*) can be categorized in various ways depending on their spatial and temporal evolution. Based on the temporal properties, these waves can either be undamped, damped or growing. And based on the spatial properties, these waves can be standing or traveling. All these three temporal possibilities are captured by the dispersion relation of the wave, which is just a relation between its frequency (ω) and wave number (k). If both ω and k are real, then the wave is undamped. If ω is complex, the wave is either growing or decaying depending on the sign of ω. It is also possible for ω to be real and k complex, but this is a very specific situation and does not usually occur naturally. Whether a wave is standing or traveling is determined by the actual expression of the perturbation.

A simple example of a growing wave is one where the resonance condition is satisfied such that the frequency of the driving perturbation is exactly same as the natural frequency of the plasma. Damping of waves is usually caused by presence of collisions. However, one of the most important examples of a damped wave is *Landau damping* [2] due to which a traveling wave excited in a plasma can get damped even in the absence of collisions. This damping takes place due to a near resonant interaction between particles and the traveling wave. Every traveling wave has a phase velocity. Particles which have velocities close to this phase velocity end up interacting with the wave thereby resulting in exchange of energy. Particles which are initially moving faster than the wave eventually slow down and those moving slower get speeded up. Essentially, all particles moving close to the wave velocity tend to start moving along with the wave which brings their velocity close to the phase velocity of the wave. Now if the plasma has more particles which are moving slower than the wave, the number of particles getting speeded up will be more than the number of particles being slowed down and hence there will be a net energy transfer from the wave to the particles thereby leading to wave damping. The reverse process can also happen if there are more particles faster than the wave thereby leading to growth of wave energy. It is important to note that this description is valid only in the linear limit of Landau damping. The full nonlinear problem has been solved by Cedric Villani for which he was awarded the Fields Medal in 2010.

Nonlinear effects of these perturbations are of immense importance and will be discussed in the next section on ponderomotive theory. In this section, we will limit ourselves to studying the linear limit. In this limit, we solve the Vlasov equation by a process known as *linearization*. The equilibrium distribution of the plasma is usually denoted by f_0 and the first-order perturbation by f_1 where it is assumed that $f_1 \ll f_0$. We do this for the electric field as well and write $E = E_0 + E_1$. Here however E_1 may not necessarily be small compared to E_0 since in most problems of interest the equilibrium field (E_0) is zero. This is specially true for the case of *quasi-neutral plasmas* where the total charge of all particles in the system is zero. Plasmas which consist of only species of charged particles are called *non-neutral* and wave phenomenon in such systems is much less understood. In this section, we limit ourselves to the case of quasi-neutral plasmas consisting of two species (ions and electrons) in equal number. Ions are usually heavy and hence assumed to form a static neutralizing background. We substitute $f = f_0 + f_1$ and $E = E_1$ in the Vlasov (Eq. (3.2)) and Poisson's equation (Eq. (3.4)) and neglect terms which contain product of the perturbed quantities (f_1, E_1),

$$\frac{\partial f_1}{\partial t} + v\frac{\partial f_0}{\partial x} + v\frac{\partial f_1}{\partial x} - \frac{eE_1}{m}\frac{\partial f_0}{\partial v} = 0 \tag{3.6}$$

$$\frac{\partial E_1}{\partial x} = -\frac{en_1}{\epsilon_0} = -\frac{e}{\epsilon_0}\int_{-\infty}^{\infty} f_1 dv \tag{3.7}$$

Separating terms containing the zero-order and first-order perturbations in Eq. (3.6), we get

$$v\frac{\partial f_0}{\partial x} = 0 \tag{3.8}$$

$$\frac{\partial f_1}{\partial t} + v\frac{\partial f_1}{\partial x} - \frac{eE_1}{m}\frac{\partial f_0}{\partial v} = 0 \tag{3.9}$$

From Eq. (3.8), it is clear that any arbitrary function of velocity (v) is a valid solution for f_0 and here we choose a Maxwellian,

$$f_0 = n_0\sqrt{\frac{m}{2\pi\kappa T}}\exp\left(-\frac{mv^2}{2\kappa T}\right) \tag{3.10}$$

In order to solve for f_1 and E_1, we do a Fourier transform of Eqs. (3.9) and (3.7) which gives,

$$-i\omega f_1 + ikvf_1 - \frac{eE_1}{m}\frac{\partial f_0}{\partial v} = 0 \tag{3.11}$$

$$ikE_1 = -\frac{e}{\epsilon_0}\int_{-\infty}^{\infty} f_1 dv \tag{3.12}$$

where $i = \sqrt{-1}$, ω is the angular frequency and k is the wave number. Substituting the expression for f_1 from (3.11) into Eq. (3.12), we get

$$ikE_1 = \frac{e^2E_1}{\epsilon_0 m}\int_{-\infty}^{\infty}\frac{\partial f_0/\partial v}{i(\omega - kv)}dv$$

$$\Rightarrow 1 + \frac{\omega_e^2}{n_0k^2}\int_{-\infty}^{\infty}\frac{\partial f_0/\partial v}{(\omega/k - v)}dv = 0 \tag{3.13}$$

where $\omega_e = \sqrt{n_0 e^2/(\epsilon_0 m)}$ is the plasma frequency. The denominator inside the integral of Eq. (3.13) becomes zero when the particle velocity (v) is equal to the phase velocity of the wave (ω/k) and this causes a problem for the integration. A method for handling this problem using complex analysis was presented by Landau in 1946 and results in the phenomenon known as Landau damping [2] described earlier.

If the phase velocity of the wave is very high, Eq. (3.13) can be approximated to the standard dispersion relation for an undamped traveling plasma wave

$$\omega^2 = \omega_e^2 + 3k^2 v_e^2 \tag{3.14}$$

where v_e is the thermal velocity. For the case of a cold plasma (zero temperature, i. e., $v_e = 0$), we get $\omega = \omega_e$ which is also known to be correct in the nonlinear limit [7]. This same dispersion relation also holds for electrons in a metal which essentially behave like a plasma.

3.4 Ponderomotive Theory

The third method of solving Vlasov equation concerns the case when the periodic perturbations in the plasma are so large that linear response theory is insufficient to account for the effects. The second order effects of these perturbations are studied under what is known as ponderomotive theory. Further higher order effects can also be important in certain plasma problems but those are usually very specific problems and not of much general interest. Once again, we are interested in finding the plasma distribution function under the effect of time-periodic spatially non-uniform electric fields. For studying the second order effects, usually the electric field is taken to be of the form

$$E(x,t) = E_s(x) + E_o(x)\cos(\omega t) \tag{3.15}$$

where E_s and E_o are arbitrary functions of x. Solving the Vlasov equation for this kind of field profile can be very tricky. That is because any method of solution for this kind of electric field has to go through a perturbation expansion and certain assumptions made about the nature of perturbation required can give very different answers. So it is very important to compare the analytical results with simulations in order to ensure correctness of the mathematical expressions.

The most straightforward method is based on the Liouville equation (Eq. (3.1)) which essentially says that the distribution function of the plasma remains constant along particle trajectories. In order to understand this, consider a point in phase space given by (x_0, v_0) and a small region around it with area $dx_0 dv_0$. Let this initial condition at $t = 0$ evolve with time according to the Newton's force equation ($F = ma$) so as to reach the point (x, v) at time t. And let the area $dx_0 dv_0$ evolve to the area $dxdv$. By Liouville theorem, the number of particles in the region $dx_0 dv_0$ will be same as that in the region $dxdv$. And if the particle motion is governed by Hamiltonian dynamics, the areas of these regions will also be the same ($dx_0dv_0 = dxdv$). Hence, it obviously follows that the distribution function at (x_0, v_0) at $t = 0$ will be the same as the distribution function at (x, v) at a future time t. Thus,

$$f(x, v, t) = f_0(x_0, v_0) \tag{3.16}$$

What we now need is a way to express (x_0, v_0) in terms of (x, v, t). If we solve the Newton's force equation for the electric field given in Eq. (3.15), we will get

$$\begin{aligned} x &= x(x_0, v_0, t) \\ v &= v(x_0, v_0, t) \end{aligned} \tag{3.17}$$

If we can invert these solutions, we will get

$$\begin{aligned} x_0 &= x_0(x, v, t) \\ v_0 &= v_0(x, v, t) \end{aligned} \tag{3.18}$$

Now simply plugging these into Eq. (3.16), we can obtain the distribution for the entire phase space at any given time. Theoretically this method looks quite clean but it is very hard to use in practice for most electric field expressions since inverting the solutions to obtain Eq. (3.18) can be very challenging. This process can however be carried out quite easily if the electric field is spatially linear, leading the following force equation

$$\ddot{x} = \frac{q}{m}\left[a + b\cos(\omega t)\right]x \tag{3.19}$$

which is the well known Mathieu equation [9] and the dot represents the time-derivative. Equation (3.19) has stable (bounded) solutions for a range of values of (a, b) outside of which it gives exponentially growing solutions. Since (3.19) is a 2nd order linear ODE, it has two linearly independent solutions which can be denoted by $\phi(t)$ and $\psi(t)$. Now, any arbitrary solution can be written as

$$\begin{aligned} x(t) &= A\phi + B\psi \\ v(t) &= A\dot{\phi} + B\dot{\psi} \end{aligned}$$

Substituting $t = 0$, we obtain

$$\begin{aligned} x_0 &= A\phi_0 + B\psi_0 \\ v_0 &= A\dot{\phi}_0 + B\dot{\psi}_0 \end{aligned}$$

which can now be easily inverted to obtain A, B in terms of x_0, v_0 which on further symbol manipulation gives x_0, v_0 in terms of x, v. Now substituting this in Eq. (3.16) gives the required plasma distribution function. It is important to note here that we can choose any arbitrary function for f_0 whose integral over phase space is finite. However, from the perspective of statistical mechanics and experimental findings, it is usually preferable to choose the Gibbs-Boltzmann distribution (Eq. (3.5)). For a general choice of the Gibbs-Boltzmann distribution for f_0, the distribution f turns out to be aperiodic (consisting of two frequencies), but for a particular choice of the spatial extent of the initial f_0, the distribution remains time-periodic for all time with the same period as the applied electric field given by Eq. (3.15) (see [4, 6] for further details).

As mentioned earlier, this method is difficult to carry out for spatially nonlinear electric fields [5, 6]. In this case, an approximate procedure known as *ponderomotive theory* [2] is widely used. Consider the equation

$$\ddot{x} = g_S(x) + g_O(x)\cos(\omega t) \tag{3.20}$$

where g_S and g_O are slowly varying functions of spatial coordinate. Figure 3.1 shows a typical trajectory of a particle governed by Eq. (3.20) and it can be seen that we can write the trajectory as a sum of a slowly varying component, $x_S(t)$ and fast oscillations, $x_O(t)$. Substituting this in Eq. (3.20), we get

$$\ddot{x}_S + \ddot{x}_O = g_S(x_S + x_O) + g_O(x_S + x_O)\cos(\omega t) \tag{3.21}$$

Since we have assumed that g_S and g_O are slowly varying functions, we can Taylor expand these functions and keep only terms up to the first order in x_O (of course, assuming that x_O is small), to get

$$\begin{aligned}\ddot{x}_S + \ddot{x}_O &= g_S(x_S) + x_O g_S'(x_S)\\ &\quad + \left[g_O(x_S) + x_O g_O'(x_S)\right]\cos(\omega t) + \mathcal{O}\left(x_O^2\right)\end{aligned} \tag{3.22}$$

We now separate the slow and fast components to get

$$\begin{aligned}\ddot{x}_S &\approx g_S(x_S) + \overline{x_O g_O'(x_S)\cos(\omega t)}\\ \ddot{x}_O &\approx x_O g_S'(x_S) + g_O(x_S)\cos(\omega t)\end{aligned} \tag{3.23}$$

where the overline represents a time-average over one period of fast oscillations, $2\pi/\omega$. In the above expression for $\ddot{x}_O$, there should also be a high frequency component of $x_O g_O'(x_S)\cos(\omega t)$, but we have neglected it since it is of a higher harmonic. If the time-scale of variation of x_O is much shorter than that of x_S, we can assume that x_S is approximately constant over one time-period of x_O. Under this approximation, we can solve for x_O to get

$$x_O = \frac{g_O(x_S)\cos(\omega t)}{-\omega^2 - g_S'(x_S)} \tag{3.24}$$

Substituting this expression of x_O in the equation for x_S in Eq. (3.23), we get

$$\ddot{x}_S \approx g_S(x_S) + \frac{1}{2}\frac{g_O(x_S)g_O'(x_S)}{-\omega^2 - g_S'(x_S)}$$

where we have used the fact that $\overline{\cos^2(\omega t)} = 0.5$ and assumed that x_S is approximately constant over one time-period of fast oscillations. This procedure is usually known as first-order averaging, and if we assume that ω^2 is much larger than $g_s'(x_s)$, we obtain an ODE for the time-evolution of the slow component, x_s,

$$\ddot{x}_s = g_s(x_s) - \frac{1}{4\omega^2}\frac{d}{dx}g_o^2(x)\Big|_{x=x_s} \tag{3.25}$$

$$\Rightarrow \frac{1}{2}v_s^2 + \phi_s(x_s) + \frac{g_o^2(x_s)}{4\omega^2} = \text{constant} \tag{3.26}$$

where ϕ_s is the potential corresponding to $g_s\ (= -d\phi_s/dx)$, the quantity $\Phi_P = \phi_s + g_o^2/4\omega^2$ is known as the *fictitious ponderomotive potential.* Since Eq. (3.25) is time-independent, the corresponding Hamiltonian given by Eq. (3.26) is be autonomous and hence one can write the effective distribution for this plasma using the Gibbs-Boltzmann distribution given by Eq. (3.5),

$$\overline{f(x,v,t)} = n_0 \exp\left[-\frac{m}{\kappa T}\left(\frac{1}{2}v^2 + \phi_s(x) + \frac{g_o^2(x)}{4\omega^2}\right)\right] \tag{3.27}$$

It can be shown using Hamiltonian averaging method [8] that this corresponds to the time-averaged distribution that would be obtained by averaging one of the infinitely many time-periodic distribution obtained by the inversion method described in the first part of this section. Now which of these solutions is the actually valid in experimental conditions is still an open question.

This idea of averaging described above is related to another very important concept in physics known as *adiabatic invariance* [23]. For time-independent forces, we know that the total energy of the system remains a constant as the particle explores the phase space. But for time-varying forces, this is not true. In many such situations, analysis can become quite simplified if there exists another quantity that is approximately constant over many oscillation-periods of the particle. For time-periodic forces of the kind given by Eq. (3.20), such an approximate constant is given by Eq. (3.26). But for more complicated situations, there is a general theory of adiabatic invariance which says that if the time-variation of the Hamiltonian is slow compared to the oscillation-period of the particle and if the time-varying component of the Hamiltonian is small in magnitude compared to the time-independent part, then there is an adiabatic invariant given by

$$I = \oint p\,dq \tag{3.28}$$

which stays approximately constant over many oscillation-periods of the particle. Here p is the generalized momentum and q is the generalized coordinate. These two quantities are related by the following Hamiltonian equation

$$\dot{q} = \frac{\partial H}{\partial p}$$
$$\dot{p} = -\frac{\partial H}{\partial q} \tag{3.29}$$

where H is the Hamiltonian of the system.

3.5 $\vec{E} \times \vec{B}$ Drift

As discussed in the previous section, it is possible to solve the Vlasov equation directly only for few specific cases of the electric-magnetic field. For many cases, we need to first analyze the single particle motion and use that to infer solutions of the Vlasov equation. The averaging method described earlier is very powerful and useful in understanding particle motion in situations where only electric field is present. However, presence of magnetic fields can further complicate matters. In this section, we take one simple case where both electric and magnetic fields are present and discuss the resulting particle trajectories.

The simplest configuration of electric-magnetic field configuration that we can consider is that of a static uniform electric field perpendicular to a static uniform magnetic field. Without loss of generality, we take the magnetic field to be in the z-direction and the electric field to be in the y-direction. If the electric field magnitude was zero, the particle would simply go along a helical path around the z-direction. And if the magnetic field magnitude was zero, the particle would keep accelerating along the y-direction (or $-y$ depending on the sign of its charge). But when we have both these fields with non-zero magnitude, the resulting particle motion can be quite different from both these possibilities. For the sake of simplicity, we assume that the particle has no initial velocity along the z-direction. Now since there are no forces along the z-direction, the particle's velocity will continue to be zero in this direction and hence its motion will be restricted to the xy-plane.

Since we have a magnetic field, it is quite likely that the particle will have an oscillatory motion around this field. Due to the presence of the electric field, this motion will not be perfectly circular but will drift in a certain direction. In order to analyze this further, let us write the equation of motion for the charged particle

$$m\frac{d\vec{v}}{dt} = q\vec{E} + q\vec{v} \times \vec{B} \tag{3.30}$$

where $\vec{E} = E_0\hat{y}$, $\vec{B} = B_0\hat{z}$, $\vec{v}$ is the particle velocity, q its charge and m its mass. We now separate the above equation into its two components along x- and y- directions

$$m\frac{dv_x}{dt} = qv_yB_0$$

$$m\frac{dv_y}{dt} = qE_0 - qv_xB_0$$

$$\Rightarrow \frac{d^2v_x}{dt^2} + \frac{q^2B_0^2}{m^2}\left(v_x - \frac{E_0}{B_0}\right) = 0$$

$$\Rightarrow v_x = \frac{E_0}{B_0} + v_o \sin\left(\frac{qB_0}{m}t + \phi\right) \tag{3.31}$$

$$\Rightarrow v_y = v_o \cos\left(\frac{qB_0}{m}t + \phi\right) \tag{3.32}$$

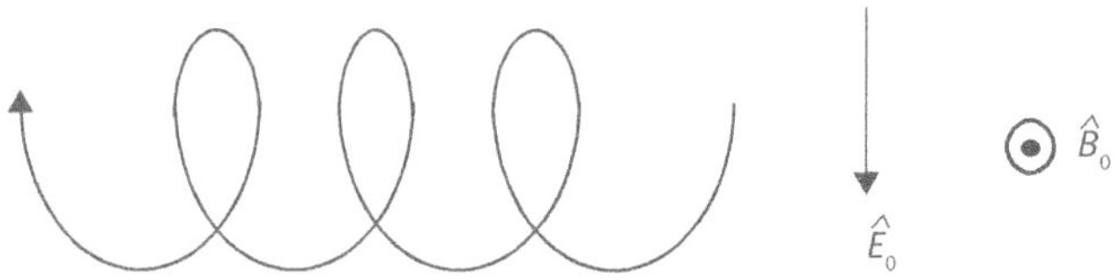

Fig. 3.2: This figure shows the $\vec{E} \times \vec{B}$ drift experienced by a charged particle.

This implies that the particle undergoes oscillatory motion around the magnetic field with the usual cyclotron frequency $\omega_c = qB_0/m$ and a constant drift of speed $v_d = E_0/B_0$ along the x-direction, as shown in the Figure 3.2. It is quite interesting to note that although the electric field is along the y-direction, the drift is only along the x-direction. This is because the electric force, qE_0, along the y-direction is exactly cancelled by the magnetic force, qv_dB_0, along the $-y$-direction. Also, the magnitude and direction of the drift is independent of the charge and mass of the particle! This is again because of the reason that the drift speed is that speed at which the electric force is exactly cancelled out by the resulting magnetic force, $qE_0 = qv_dB_0$.

Now we may ask what happens when the electric and magnetic fields are not perpendicular to each other. One trivial case is when the electric and magnetic fields are parallel, i. e., both along the z-direction. In this case, the particle motion is again given by a helix but where the particle keeps accelerating along the z or $-z$-direction depending on its charge. The spiral motion around the magnetic field lines is independent of this axial motion and hence the radius and period of oscillations remains unchanged.

If the electric and magnetic fields are neither parallel nor perpendicular, the z-component of velocity is again non-zero and we need to solve the full 3D equations of motion with all the three components of velocity. However, this case is also analytically tractable and very similar to the case when the electric and magnetic fields are perpendicular to each other. Without loss of generality, we can again take the magnetic field to be along the z-direction and the electric field to be in the yz-plane. The equations and solutions for the x- and y-components of the velocity are again given by Eqs. (3.31) and (3.32). The equation for v_z is given by $mdv_z/dt = qE_z$ which can be easily solved to give $v_z(t) = v_{z0} + qE_zt/m$, since E_z is a constant in our case. Thus, for a general electric field, $\vec{E}$, and a magnetic field, $\vec{B}$, the drift velocity is given by

$$\vec{v}_d = \frac{\vec{E} \times \vec{B}}{|\vec{B}|^2} \tag{3.33}$$

4 Plasma Fluid Theory

4.1 Introduction

In the previous chapter, we have studied the plasma kinetic theory and seen how to solve for the plasma distribution function in various configurations of electric field. The 1D plasma distribution function is a function of three independent variables (position, velocity, time) and contains lot of information about the state of the plasma at any given time. However, in many practical situations it is very difficult to solve the Vlasov equation in order to get the distribution function. In order to make life easier, what can be done is to write an equation for the moments of the distribution function (moments as in integrals over velocity) which turn out to be lot simpler to solve. These set of equations are known as the *fluid equations* since now instead of considering the plasma as being composed of a collection of particles, it is considered to be made of various fluid components (usually one component for each charged particle species). The difference between particles and fluids is very important. Particles are discrete entities which exist at a point in phase space. Fluids on the other hand are spread out over a continuous range of configuration space (or position space) and can have different velocities at different spatial points. So in some sense, a fluid description captures the average of the particle behavior due to which it obviously misses out some of the effects that are seen only in the particle (or discrete or kinetic) picture. One example of such an effect is Landau damping (described in the previous chapter) which cannot be recovered from the collisionless fluid equations. So, for a given problem, it is important to know apriori whether kinetic effects are going to be important in the phenomenon of interest. If kinetic effects are going to be important, it is imperative to solve for the plasma distribution function. However, if kinetic effects can be ignored and only the averaged fluid-like behavior is of interest, then it is sufficient to solve the fluid equations. Here it is important to note that solving the fluid equations is also not a trivial job. Just that it is easier than solving the Vlasov equations in many problems. There are also several problems where solving the fluid equations analytically is also very tedious and recourse must be made to computer simulations. One example of such a problem is turbulence which is considered to be one the toughest challenges in physics.

4.2 Derivation of the Fluid Equations

In this section we will derive the 1D fluid equations starting from the Vlasov equation (Eq. (3.2)) and this can be generalized to higher dimensions. As mentioned earlier, fluid equations are equations for the moments (with respect to velocity) of the plasma distribution function. Now of course there can be infinitely many such moments and

https://doi.org/10.1515/9783110570038-055

thereby an infinite number of equations. However, we restrict ourselves to the first two moments which serve the purpose for most plasma problems of practical interest. These first two moments are the plasma density and fluid velocity. In some problems a third moment is also studied and is the plasma energy.

The first moment of the distribution function [density, $n(x,t)$] is obtained simply by integrating $f(x,v,t)$ over all velocity,

$$n(x,t) = \int_{-\infty}^{\infty} f(x,v,t)\,dv \tag{4.1}$$

And similarly, the second moment [plasma velocity, $u(x,t)$] is given by,

$$u(x,t) = \frac{1}{n(x,t)} \int_{-\infty}^{\infty} vf dv \tag{4.2}$$

Now to find an equation for $n(x,t)$ and $u(x,t)$, we take moments of the Vlasov equation in the same fashion as Eqs. (4.1) and (4.2). As described in the previous chapter, the Vlasov equation is given by

$$\frac{\partial f}{\partial t} + v\frac{\partial f}{\partial x} + \frac{qE(x,t)}{m}\frac{\partial f}{\partial v} = 0 \tag{4.3}$$

where $E(x,t)$ is the electric field, q is the charge of particles and m their mass. Here we consider only the 1D Vlasov equation but the procedure described below can be generalized to the 3D case including the magnetic field in the force term. In order to find the first fluid equation, we simply integrate Eq. (4.3) with respect to velocity,

$$\int_{-\infty}^{\infty} dv \left[\frac{\partial f}{\partial t} + v\frac{\partial f}{\partial x} + \frac{qE(x,t)}{m}\frac{\partial f}{\partial v}\right] = 0$$

$$\Rightarrow \frac{\partial}{\partial t}\left(\int_{-\infty}^{\infty} fdv\right) + \frac{\partial}{\partial x}\left(\int_{-\infty}^{\infty} vfdv\right) + \frac{qE(x,t)}{m}\left(\int_{-\infty}^{\infty} \frac{\partial f}{\partial v}dv\right) = 0 \tag{4.4}$$

since the velocity variable v is independent of (x,t) and can be taken inside the derivatives. Integrating the above equation and using Eqs. (4.1) and (4.2) we get the plasma *continuity equation*,

$$\frac{\partial n}{\partial t} + \frac{\partial nu}{\partial x} = 0 \tag{4.5}$$

where we have assumed that f decays to zero as $v \to \pm\infty$, which is quite reasonable for any bounded plasma. In order to obtain the second fluid equation, we multiply Eq. (4.3) with v and then integrate with respect to the same variable v,

$$\int_{-\infty}^{\infty} dv \left[v\frac{\partial f}{\partial t} + v^2\frac{\partial f}{\partial x} + \frac{qvE(x,t)}{m}\frac{\partial f}{\partial v}\right] = 0$$

$$\Rightarrow \frac{\partial}{\partial t}\left(\int_{-\infty}^{\infty} vfdv\right) + \frac{\partial}{\partial x}\left(\int_{-\infty}^{\infty} v^2 fdv\right) + \frac{qE(x,t)}{m}\left(\int_{-\infty}^{\infty} v\frac{\partial f}{\partial v}dv\right) = 0 \tag{4.6}$$

The first term in the above equation gives $\partial(nu)/\partial t$ and the third term can be integrated using integration by parts to give $-qnE/m$. It is the second term which is a little tricky. The fluid velocity u given by Eq. (4.2) is actually the mean of the plasma distribution function. Using this information, we can write the second term of Eq. (4.6) as

$$\begin{aligned}
&\frac{\partial}{\partial x}\left(\int_{-\infty}^{\infty} v^2 f dv\right)\\
&= \frac{\partial}{\partial x}\left(\int_{-\infty}^{\infty} (v-u)^2 f dv + \int_{-\infty}^{\infty} u^2 f dv\right)\\
&= \frac{\partial}{\partial x}\left(\int_{-\infty}^{\infty} (v-u)^2 f dv\right) + \frac{\partial n u^2}{\partial x}\\
&= \frac{1}{m}\frac{\partial P}{\partial x} + \frac{\partial n u^2}{\partial x} \qquad (4.7)
\end{aligned}$$

where P is the pressure term. Substituting the corresponding expressions in the three terms of Eq. (4.6), we get

$$\begin{aligned}
&\frac{\partial nu}{\partial t} + \left(\frac{1}{m}\frac{\partial P}{\partial x} + \frac{\partial n u^2}{\partial x}\right) - \frac{qnE(x,t)}{m} = 0\\
&\Rightarrow n\frac{\partial u}{\partial t} + u\frac{\partial n}{\partial t} + u\frac{\partial nu}{\partial x} + nu\frac{\partial u}{\partial x} = \frac{qnE(x,t)}{m} - \frac{1}{m}\frac{\partial P}{\partial x} \qquad (4.8)
\end{aligned}$$

Now we can use the continuity equation (Eq. (4.5)) to simplify the above equation to finally get the plasma fluid *momentum equation*,

$$mn\frac{\partial u}{\partial t} + mnu\frac{\partial u}{\partial x} = qnE(x,t) - \frac{\partial P}{\partial x} \qquad (4.9)$$

In order to solve the above equation, we need an expression for the electric field (E) pressure term (P) and its dependence on n and u. The electric field is simply given by the Poisson equation,

$$\frac{\partial E}{\partial x} = \frac{qn}{\epsilon_0} \qquad (4.10)$$

Finding the precise expression for the pressure is quite challenging and for most practical purposes, this is usually done by using thermodynamic considerations. Pressure is known to depend on the temperature through the equation of state

$$P = nRT \qquad (4.11)$$

where $R = 8.314\ \mathrm{J\,mol^{-1}\,K^{-1}}$ is the gas constant and T is the plasma temperature. For an isothermal plasma, we have $\partial_x P = RT\partial_x n$ and for adiabatic behavior, we have $\partial_x P = \gamma RT\partial_x n$ where γ is the ratio of specific heats. For a 1D plasma, $\gamma = 3$. Thus, the final set of 1D plasma fluid equations are

$$\frac{\partial n}{\partial t} + \frac{\partial nu}{\partial x} = 0 \qquad \text{Continuity Equation}$$

$$mn\frac{\partial u}{\partial t} + mnu\frac{\partial u}{\partial x} = qnE(x,t) - \frac{\partial P}{\partial x} \qquad \text{Momentum Equation}$$

$$\frac{\partial E}{\partial x} = \frac{qn}{\epsilon_0} \quad \text{Poisson's Equation}$$
$$P = nRT \quad \text{Ideal Gas Equation} \tag{4.12}$$

For the 2D or 3D case, the momentum equation will have an additional $\vec{u} \times \vec{B}$ force term along with $\vec{E}$ and the Poisson equation has to be replaced by the entire set of Maxwell's equations described in the first chapter of this book. Also, the spatial derivative has to be replaced by the gradient or divergence operator as the case may be.

It is also important to note that the above set of plasma fluid equations are to be used to each species present in the plasma. Hence if we have a quasi-neutral plasma consisting of ions and electrons, we will have two sets of fluid equations for each species making the total number of equations to be seven (two continuity equations, two momentum equations, two pressure equations and one Poisson's equations). The Poisson's equation always remains one since the electric field is determined collectively by all the charged particle species present. In the next section, we will see how to solve the plasma fluid equations in order to obtain the dispersion relations for the electrostatic waves (also known as *Langmuir waves*). There are many other kinds of plasma waves that exist but are not very relevant to plasmonics and have been ignored in this book.

4.3 Electrostatic Wave

The term *electrostatic wave* appears to be a misnomer to begin with! This is because electrostatics is the study of temporally constant electric fields. We know from Maxwell's equations that usually a time-variation of the electric field produces a time-varying magnetic field thereby rendering the phenomenon to be *electromagnetic* instead of electrostatic. However, in the case of 1D plasmas it is possible for a time-varying electric field to produce no time-varying magnetic field. In the Maxwell's equations, there are two equations that connect the electric and magnetic field

$$\vec{\nabla} \times \vec{B} = \mu_0 \vec{J} + \mu_0 \epsilon_0 \frac{\partial \vec{E}}{\partial t} \tag{4.13}$$

$$\vec{\nabla} \times \vec{E} = -\frac{\partial \vec{B}}{\partial t} \tag{4.14}$$

where $\vec{J} = qn\vec{u}$ is the current density. For a 1D plasma, $\vec{\nabla} \times \vec{E}$ is obviously zero since $\vec{E}$ is in the x-direction and it also varies only in that direction. It can also be shown that the RHS of (4.13) is also zero. These two results together imply that the magnetic field is zero for such plasmas. Taking the time-derivative of the Poisson's equation and using the continuity equation, we get

$$\frac{\partial^2 E}{\partial x \partial t} = \frac{q}{\epsilon_0} \frac{\partial n}{\partial t}$$

$$\frac{\partial^2 E}{\partial x \partial t} = -\frac{q}{\epsilon_0}\frac{\partial(nu)}{\partial x}$$
$$= -\frac{1}{\epsilon_0}\frac{\partial J}{\partial x}$$
$$\Rightarrow \frac{\partial}{\partial x}\left[\epsilon_0 \frac{\partial E}{\partial t} + J\right] = 0$$

Thus $J + \epsilon_0 \partial E / \partial t$ is at most a constant which can be taken to be zero. Hence, there exists 1D plasma waves that are electrostatic in nature! In fact, we had obtained the dispersion relation for these waves using the Vlasov equation in Section (3.3) of Chapter 3. Here we will derive the dispersion relation for the same wave using the fluid equations. As will be seen, the fluid derivation is much simpler but unlike the Vlasov derivation does not give any information regarding *Landau damping*.

In order to find the dispersion relation for the electrostatic waves (also called *Langmuir waves*) in a two-component plasma (ions and electrons) using fluid equations, we carry out the same *linearization* procedure described in Section (3.3) of Chapter 3. We assume that both the ion and electron fluid of the unperturbed plasma has spatially uniform density (n_0 = constant), zero fluid velocity and zero electric field. The ions are heavy and are assumed to form a neutralizing static background. Hence, its only the electron fluid which experiences the perturbation. Thus, for the the electron fluid, we write $n = n_0 + n_1(x, t)$, $u = u_1(x, t)$ and $E = E_1(x, t)$. Substituting this in the fluid equations (Eq. (4.12)) and keeping only the linear terms, we get,

$$\begin{aligned}
\frac{\partial n_1}{\partial t} + n_0 \frac{\partial u_1}{\partial x} &= 0 \\
m_e n_0 \frac{\partial u_1}{\partial t} &= -e n_0 E_1 - \frac{\partial P}{\partial x} \\
\frac{\partial E_1}{\partial x} &= \frac{e(n_i - n_e)}{\epsilon_0} = -\frac{e n_1}{\epsilon_0} \\
P &= nRT
\end{aligned} \tag{4.15}$$

where e is the magnitude of electron charge and m_e the electron mass. In order to solve the above equation, we assume that our wave propagation is adiabatic in nature. Thus, substituting $\partial_x P = 3RT\partial_x n$ in the above equation and taking a Fourier transform in both the spatial and temporal coordinates, we get,

$$\begin{aligned}
-i\omega n_1 + ikn_0 u_1 &= 0 \\
-i\omega m_e n_0 u_1 &= -e n_0 E_1 - 3ikRTn_1 \\
ikE_1 &= \frac{e(n_i - n_e)}{\epsilon_0} = -\frac{e n_1}{\epsilon_0}
\end{aligned} \tag{4.16}$$

Eliminating n_1, u_1, E_1 from the above equations, we get

$$\omega^2 = \omega_e^2 + 3k^2 v_e^2 \tag{4.17}$$

where $\omega_e = \sqrt{n_o e^2/(m_e \epsilon_0)}$ is the electron plasma frequency and $v_e = \sqrt{RT/m_e}$ the electron thermal velocity. If the wavelength of the waves under consideration is very high $(k = 2\pi/\lambda \to 0)$, then we can drop the second term on the RHS of Eq. (4.17) to get

$$\omega = \pm\omega_e \tag{4.18}$$

which is same as the natural frequency of oscillations of free electrons in a metal.

4.4 Plasma Conductivity and Permittivity

In this section, we shall derive the expressions for conductivity and permittivity of a plasma and discuss its similarity with these relations for a metal. Using Eq. (4.16), we can write

$$\begin{aligned} -i\omega m_e n_0 u_1 &= -en_0 E_1 + 3ikRT\frac{ik\epsilon_0}{e}E_1 \\ \Rightarrow u_1 &= -i\left(\frac{e}{\omega m_e} + \frac{3k^2\epsilon_0 RT}{e\omega m_e n_0}\right)E_1 \end{aligned}$$

So, the plasma electron current density is given by

$$\begin{aligned} J_1 &= -en_0 u_1 \\ &= ien_0\left(\frac{e}{\omega m_e} + \frac{3k^2\epsilon_0 RT}{e\omega m_e n_0}\right)E_1 \\ &= i\left(\frac{e^2 n_0}{\omega m_e} + \frac{3k^2\epsilon_0 RT}{\omega m_e}\right)E_1 \end{aligned}$$

From Ohm's law, we know that $J = \sigma E$ and hence the conductivity for the plasma electron fluid is given by

$$\sigma = i\left(\frac{e^2 n_0}{\omega m_e} + \frac{3k^2\epsilon_0 RT}{\omega m_e}\right) \tag{4.19}$$

We know from Eq. (2.8) of Chapter 2 that the permittivity of a medium is related to its conductivity by the relation

$$\epsilon = \epsilon_0 + i\frac{\sigma}{\omega} \tag{4.20}$$

Substituting Eq. (4.19) in the above equation, we get,

$$\epsilon = \epsilon_0 - \left(\frac{e^2 n_0}{\omega^2 m_e} + \frac{3k^2\epsilon_0 RT}{\omega^2 m_e}\right) \tag{4.21}$$

If the wavelength of the Langmuir waves is very long ($k \to 0$) or the temperature is very low, the last term in the above equation can be neglected and we obtain the plasma dielectric constant

$$\frac{\epsilon}{\epsilon_0} = 1 - \frac{\omega_e^2}{\omega^2} \tag{4.22}$$

As shown in Eq. (2.11), the dielectric constant for metals in a certain frequency range is exactly same as that of plasma electrons given by Eq. (4.22) and that is the relation between *plasmonics* and *plasmas*. The only difference is the actual value of the $\omega_e = \sqrt{e^2 n_0 / (m_e \epsilon_0)}$. In the expression for ω_e, the only variable is the electron density since all other parameters are physical constants. For a terrestrial plasma, the value of n_0 can vary widely from 10^7 m^{-3} to 10^{32} m^{-3} and hence the plasma frequency can approximately range from 30 kHz to 10 PHz. For typical metals, however, the density of electrons is roughly around 10^{29} m^{-3} thereby leading to a plasma frequency in the ultraviolet frequency range (~ 3 PHz). This implies that at optical frequencies (430–770 THz), the permittivity of metals is negative due to which most metals reflect visible light giving them a shiny texture. This negative value of permittivity is also essential for the existence of surface plasmons as we will see in later chapters. At frequencies much lower than optical frequencies, the permittivity becomes very highly negative and the metals behave as perfect electric conductor (PEC) thereby flushing out all electric and magnetic fields from their interiors.

4.5 Electromagnetic Waves

In the previous section, we derived the expression for permittivity of a plasma using the linearized fluid equations for an electrostatic wave. However, what we are truly interested in is an electromagnetic wave. And it is very much possible that the permittivity for electromagnetic waves is different from that of electrostatic waves. In this section, we will derive the dispersion relation for an electromagnetic wave in a plasma and find that the corresponding permittivity turns out to be exact same as given by Eq. (4.22). Here it must be noted that electromagnetic waves cannot exist in a purely 1D system and so we must consider the 3D fluid equations coupled with the full set of Maxwell's equations (unlike the electrostatic case where only the Poisson equation is sufficient).

$$
\begin{aligned}
\frac{\partial n_s}{\partial t} + \vec{\nabla} \cdot \left(n_s \vec{u}_s\right) &= 0 \\
m_s n_s \frac{\partial \vec{u}_s}{\partial t} + m n_s \left(\vec{u}_s \cdot \vec{\nabla}\right) \vec{u}_s &= q n_s \left(\vec{E} + \vec{u}_s \times \vec{B}\right) - \vec{\nabla} P_s \\
P_s &= n_s R T_s \\
\vec{\nabla} \times \vec{B} &= \mu_0 \vec{J} + \mu_0 \epsilon_0 \frac{\partial \vec{E}}{\partial t} \\
\vec{\nabla} \times \vec{E} &= -\frac{\partial \vec{B}}{\partial t} \\
\vec{\nabla} \cdot \vec{E} &= \frac{e\left(n_i - n_e\right)}{\epsilon_0} \\
\vec{\nabla} \cdot \vec{B} &= 0
\end{aligned}
\tag{4.23}
$$

where the subscript s can be either ions (i) or electrons (e). Hence, we have four fluid equations (2 for each species), two ideal gas equations (one for each species) and four Maxwell's equations (common for both species). In order to simplify our analysis, we study only transverse waves since longitudinal waves are electrostatic in nature and follow the analysis presented in Section (4.3). We also consider only the case of unmagnetized plasmas (no externally applied zeroth order magnetic field). Magnetized plasmas show a very rich behavior with the presence of several different kinds of plasma waves and the reader is encouraged to go through these in detail from other textbooks on plasma physics [2].

Since we are considering only transverse waves, we have $\vec{k} \cdot \vec{E} = 0 = \vec{k} \cdot \vec{B}$. Substituting this in Poisson's equation, we get $n_i = n_e$ and hence the density of electrons remains constant (since we again assume the ions to form a neutralizing static background). If $n_e = n_0$ is a constant, from the continuity equation, we obtain $\vec{k} \cdot \vec{u}_e = 0$ which again implies that the second term on LHS of the momentum equation is zero. If the electron density is constant and if we assume isothermal conditions, the pressure term in the momentum equation also becomes zero. Linearizing the remaining non-zero terms in the above equations for electron motion, we get

$$m_e n_0 \frac{\partial \vec{u}_1}{\partial t} = -e n_0 \vec{E}_1 \tag{4.24}$$

$$\vec{\nabla} \times \vec{B}_1 = -\mu_0 e n_0 \vec{u}_1 + \mu_0 \epsilon_0 \frac{\partial \vec{E}_1}{\partial t} \tag{4.25}$$

$$\vec{\nabla} \times \vec{E}_1 = -\frac{\partial \vec{B}_1}{\partial t} \tag{4.26}$$

where the subscript 1 refers to the first order perturbed quantities (same notation as used in Section (4.3)). Note that the second order $\vec{u}_1 \times \vec{B}_1$ term has been dropped in the momentum equation since we are considering only linear electromagnetic waves and the plasma has no zeroth order velocity or magnetic field.

The next step is to eliminate $\vec{u}_1$ and $\vec{B}_1$ from the above equations. In order to do this, we take the time derivative of Eq. (4.25) and the curl or Eq. (4.26) to get

$$\vec{\nabla} \times \left(\vec{\nabla} \times \vec{E}_1\right) = -\mu_0 e n_0 \frac{e\vec{E}_1}{m_e} - \mu_0 \epsilon_0 \frac{\partial^2 \vec{E}_1}{\partial t^2} \tag{4.27}$$

Taking spatial and temporal Fourier transform of the above equation (and assuming plane waves, $\vec{\nabla} \times \vec{\nabla} \times \rightarrow k^2$), we get,

$$k^2 = -\mu_0 \frac{e^2 n_0}{m_e} + \omega^2 \mu_0 \epsilon_0$$

which implies

$$\omega^2 = \omega_e^2 + c^2 k^2 \tag{4.28}$$

where $\omega_e = \sqrt{e^2 n_0 / (m_e \epsilon_0)}$ is the plasma frequency and $c = \sqrt{1/(\mu_0 \epsilon_0)}$ is the speed of light. Equation (4.28) is the dispersion relation for a linear transverse electromagnetic wave in a plasma.

Now we will use the above dispersion relation to find the permittivity of the plasma. We know that the refractive index of a material is given by

$$n = \frac{1}{\sqrt{\mu_r \epsilon_r}} = \frac{ck}{\omega} \tag{4.29}$$

where μ_r is the relative permeability and ϵ_r is the relative permittivity. Substituting the expression for k from Eq. (4.28) in the above, we get

$$\epsilon_r = \frac{\epsilon}{\epsilon_0} = 1 - \frac{\omega_e^2}{\omega^2} \tag{4.30}$$

where μ_r is assumed to be unity since the plasma is a non-magnetic material. It can be seen that Eq. (4.30) is exactly same as the permittivity expression (Eq. (4.22)) derived from the electrostatic wave dispersion relation.

For the propagation of EM waves, Eq. (4.30) gives a very crucial bound on the frequencies of electromagnetic waves which can propagate in a plasma. For $\omega < \omega_e$, the permittivity becomes negative and hence it prevents electromagnetic waves from propagating inside it. So, if an electromagnetic wave with frequency ω is incident on a plasma with electron plasma frequency $\omega_e > \omega$, the wave will be reflected back and will exponentially decay inside the plasma boundary without being transmitted. This is similar to the skin effect we see for usual metallic conductors. This property is very crucial in many plasma experiments where this decaying behavior is responsible for plasma heating which is sometimes desirable and sometimes not. As discussed earlier, this negative permittivity is also very crucial for the existence of surface plasmons which are surface electromagnetic waves that travel on the surface of a metal and dielectric.

5 Surface Plasmon Polaritons (SPP)

Surface Plasmon Polaritons (SPP) are electromagnetic waves traveling on the interface of a metal and dielectric. In our analysis, we will assume the dielectric to have a constant permittivity but the permittivity of the metal will be taken to depend on frequency as given by Eqs. (2.11) or (4.30),

$$\epsilon_r = \frac{\epsilon}{\epsilon_0} = 1 - \frac{\omega_p^2}{\omega^2} \tag{5.1}$$

where ω_p is the plasma frequency of the metal electrons and ϵ_r is the relative permittivity. We will see in later sections why this is important. It is important to note that the above expression for permittivity is valid only in a frequency range around ω_p. For all frequencies, the permittivity is given by Eq. (2.5) from which the above expression has also been derived. For simplicity, the permeability of both dielectric and metal is assumed to be constant and same as that of free space. The metal permittivity is taken to be spatially uniform although it varies with time/frequency.

In order to find the expressions of the electric and magnetic fields comprising the SPP waves, we need to solve the Maxwell's equations in matter given by Eq. (1.29) coupled with the boundary conditions given by Eq. (1.33),

$$\vec{\nabla} \cdot \vec{D} = \rho_f \tag{5.2}$$

$$\vec{\nabla} \times \vec{E} = -\frac{\partial \vec{B}}{\partial t} \tag{5.3}$$

$$\vec{\nabla} \cdot \vec{B} = 0 \tag{5.4}$$

$$\vec{\nabla} \times \vec{H} = \vec{J}_f + \frac{\partial \vec{D}}{\partial t} \tag{5.5}$$

$$D_1^{\perp} - D_2^{\perp} = \sigma_f \tag{5.6}$$

$$B_1^{\perp} - B_2^{\perp} = 0 \tag{5.7}$$

$$\vec{E}_1^{\parallel} - \vec{E}_2^{\parallel} = 0 \tag{5.8}$$

$$\vec{H}_1^{\parallel} - \vec{H}_2^{\parallel} = \vec{K}_f \times \vec{n} \tag{5.9}$$

Assuming all free charges and currents to be zero, taking a curl of Eq. (5.3) and substituting Eq. (5.5), we get

$$\vec{\nabla} \times \left(\vec{\nabla} \times \vec{E}\right) = -\mu_0 \frac{\partial^2 \vec{D}}{\partial t^2}$$

$$\Rightarrow \vec{\nabla}^2 \vec{E} = \mu_0 \frac{\partial^2 \vec{D}}{\partial t^2}$$

since we have assumed the permittivity of both metal and dielectric to be spatially uniform. It is important to note that the RHS of the above equation cannot be written as $-\mu_0 \epsilon \partial^2 \vec{E} / \partial t^2$ since the permittivity, ϵ, depends on time/frequency in our case and

https://doi.org/10.1515/9783110570038-065

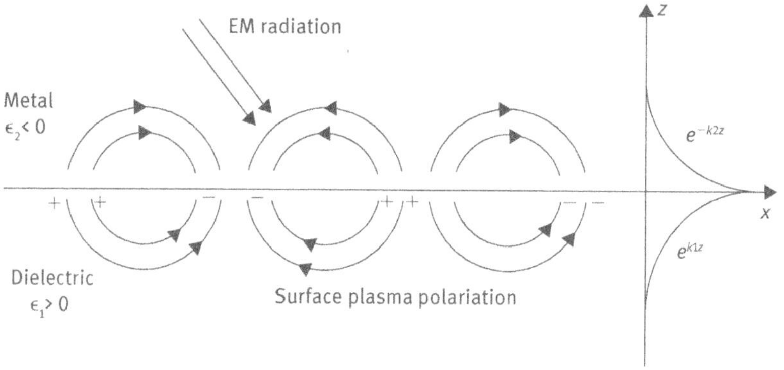

Fig. 5.1: The geometry that supports a surface plasmon wave.

cannot be taken out of the derivative. What we need to do is take a Fourier transform of the above equation in time-domain $(\partial/\partial t \to -i\omega)$, in which case the derivative becomes a multiplication, giving

$$\vec{\nabla}^2 \vec{E} + \mu_0 \epsilon \omega^2 \vec{E} = 0 \tag{5.10}$$

which is known as the *Helmholtz equation*. In order to make further progress, we need to define the geometry as shown in Figure 5.1. Without loss of generality, we assume $z = 0$ to be the interface between the metal $(z > 0)$ and dielectric $(z \le 0)$ and the wave to be propagating along the $+x$ direction. Thus, there is no variation along the y-direction and the electric field can be written as $E(x, y, z) = E(z)\, e^{i\beta x}$, where β is the propagation constant. Substituting this in Eq. (5.10), we get

$$\frac{\partial^2 \vec{E}(z)}{\partial z^2} + \left(\mu_0 \epsilon \omega^2 - \beta^2\right) \vec{E}(z) = 0 \tag{5.11}$$

and a similar equation can be written for the magnetic field. The above equation is quite widely used in the analysis of electromagnetic waves in various scenarios.

5.1 SPP on Single Interface

In order to solve for the SPP wave expressions, it is not enough to solve the Helmholtz equation since we also need the inter-relationship between the electric and magnetic field. This is important for satisfying the boundary conditions in a consistent manner. We take a spatial and temporal Fourier transform $(\partial/\partial x \to i\beta, \partial/\partial t \to -i\omega)$ of Eqs. (5.3) and (5.5) to get six equations for all the components of the electric and magnetic field

$$\frac{\partial E_y}{\partial z} = -i\omega\mu_0 H_x$$

$$\begin{aligned}\frac{\partial E_x}{\partial z} - i\beta E_z &= i\omega\mu_0 H_y \\ i\beta E_y &= i\omega\mu_0 H_z \\ \frac{\partial H_y}{\partial z} &= i\omega\epsilon E_x \\ \frac{\partial H_x}{\partial z} - i\beta H_z &= -i\omega\epsilon E_y \\ i\beta H_y &= -i\omega\epsilon E_z \end{aligned} \tag{5.12}$$

The above set of 6 equations can be separated into two independent sets of 3 equations each. One set called the TM (transverse magnetic) mode consists of only E_z, E_x and H_y. And the other set called TE (transverse electric) mode consists of only H_x, H_z and E_y. These modes are so called since in the TM mode, the magnetic field is perpendicular to the direction of wave propagation ($+x$) and in the TE mode, the electric field is perpendicular to the direction of propagation. In some books/papers, the TM mode is also called TM_z mode since in this mode the magnetic field is perpendicular to the z-direction. Similarly, TE mode is sometimes referred to as the TE_z mode. We will now analyze these two modes separately.

5.1.1 TE Mode of SPP

The equations for the TE mode are

$$\begin{aligned}\frac{\partial E_y}{\partial z} &= -i\omega\mu_0 H_x \\ i\beta E_y &= i\omega\mu_0 H_z \\ \frac{\partial H_x}{\partial z} - i\beta H_z &= -i\omega\epsilon E_y \end{aligned} \tag{5.13}$$

In order to solve for the expressions of each field component, we need to write them separately for the two regions (dielectric and metal) and then use boundary conditions given by Eqs. (5.6) to (5.9). We begin by writing an expression for E_y and then derive H_x and H_z from E_y using the above equation,

$$E_y = \begin{cases} A_1 e^{i\beta x} e^{k_1 z}, & z \le 0 \\ A_2 e^{i\beta x} e^{-k_2 z}, & z > 0 \end{cases} \tag{5.14}$$

where $k_1, k_2 > 0$ are the decay constants in the two regions. Substituting this expression for E_y in Eq. (5.13), we get

$$H_x = \begin{cases} (iA_1 k_1/\omega\mu_0)\, e^{i\beta x} e^{k_1 z}, & z \le 0 \\ -(iA_2 k_2/\omega\mu_0)\, e^{i\beta x} e^{-k_2 z}, & z > 0 \end{cases} \tag{5.15}$$

and

$$H_z = \begin{cases} (A_1\beta/\omega\mu_0)\, e^{i\beta x} e^{k_1 z}, & z \le 0 \\ (A_2\beta/\omega\mu_0)\, e^{i\beta x} e^{-k_2 z}, & z > 0 \end{cases} \tag{5.16}$$

where $\epsilon_1 > 0$ is the permittivity of the dielectric and ϵ_2 of the metal. Since we have assumed free charges and free currents to be zero, Eqs. (5.8) and (5.9) imply that E_y and H_x are equal on both sides of the metal-dielectric interface at $z = 0$. Hence, Eq. (5.14) implies $A_1 = A_2$. And Eq. (5.15) implies $A_1 k_1 + A_2 k_2 = 0$. Taken together, we get

$$A_1 (k_1 + k_2) = 0 \tag{5.17}$$

But since $k_1, k_2 > 0$, the only possible solution to this equation is $A_1 = 0 = A_2$ which implies the absence of a surface plasmon wave. Hence, the surface plasmon wave on a single metal-dielectric interface cannot be of TE mode. Now let us analyze the TM mode and see if this can exist.

5.1.2 TM Mode of SPP

The equations for the TM mode are

$$\begin{aligned} \frac{\partial E_x}{\partial z} - i\beta E_z &= i\omega\mu_0 H_y \\ \frac{\partial H_y}{\partial z} &= i\omega\epsilon E_x \\ i\beta H_y &= -i\omega\epsilon E_z \end{aligned} \tag{5.18}$$

We proceed in the same manner as we did for the TE mode by writing the expressions for the three components on both sides of the interface. The only difference is that for the TM case we begin with H_y instead of E_y,

$$H_y = \begin{cases} A_1 e^{i\beta x} e^{k_1 z}, & z \le 0 \\ A_2 e^{i\beta x} e^{-k_2 z}, & z > 0 \end{cases} \tag{5.19}$$

where $k_1, k_2 > 0$ are the decay constants in the two regions and the z-dependence has been chosen to ensure that the wave decays as we move away from the interface at $z = 0$. Substituting this expression for H_y in Eq. (5.18), we get

$$E_x = \begin{cases} -(iA_1 k_1/\omega\epsilon_1)\, e^{i\beta x} e^{k_1 z}, & z \le 0 \\ (iA_2 k_2/\omega\epsilon_2)\, e^{i\beta x} e^{-k_2 z}, & z > 0 \end{cases} \tag{5.20}$$

and

$$E_z = \begin{cases} -(A_1\beta/\omega\epsilon_1)\, e^{i\beta x} e^{k_1 z}, & z \le 0 \\ -(A_2\beta/\omega\epsilon_2)\, e^{i\beta x} e^{-k_2 z}, & z > 0 \end{cases} \tag{5.21}$$

From the boundary condition given by Eq. (5.9), we get that H_y must be equal on both sides of the interface thereby implying $A_1 = A_2$. Same result is obtained from Eq. (5.6) which implies continuity of $\epsilon_i E_z$. Equation (5.8) implies continuity of E_x and using this in Eq. (5.20), we get

$$\frac{k_1}{\epsilon_1} = -\frac{k_2}{\epsilon_2} \tag{5.22}$$

where we have used the fact that $A_1 = A_2$. Equation (5.22) is very important. Since $k_1, k_2 > 0$, this equation can be satisfied only if ϵ_1 and ϵ_2 are of opposite signs. And this is where the importance of the dispersion relation given by Eq. (5.1) comes in. According to this dispersion relation, the permittivity of the metal is negative if $\omega \lesssim \omega_p$.

In order to proceed further, we note that H_y must also satisfy the wave equation given by Eq. (5.11). Thus, substituting Eq. (5.19) in Eq. (5.11), we get

$$\begin{aligned} k_1^2 &= \beta^2 - \omega^2 \mu_0 \epsilon_1 \\ k_2^2 &= \beta^2 - \omega^2 \mu_0 \epsilon_2 \end{aligned} \tag{5.23}$$

Substituting Eq. (5.22) in the above equation, we get an expression for the propagation constant,

$$\beta = k_0 \sqrt{\frac{\epsilon_{r1}\epsilon_{r2}}{\epsilon_{r1} + \epsilon_{r2}}} \tag{5.24}$$

where $k_0 = \omega\sqrt{\mu_0 \epsilon_0}$ is the propagation constant in free space, $\epsilon_{r1} = \epsilon_1/\epsilon_0$ and $\epsilon_{r2} = \epsilon_2/\epsilon_0$. We have already seen earlier that ϵ_1 and ϵ_2 must be of the opposite sign. Using this fact, Eq. (5.24) implies that $\epsilon_{r1} + \epsilon_{r2} < 0$ in order to ensure that β is real (and not imaginary), which is required for the SPP wave to be propagating. Substituting this condition in the metal dispersion relation, Eq. (5.1), we get

$$\begin{aligned} &1 - \frac{\omega_p}{\omega^2} + \epsilon_{r1} < 0 \\ &\Rightarrow \omega < \frac{\omega_p}{\sqrt{1 + \epsilon_{r1}}} = \omega_{SP} \end{aligned} \tag{5.25}$$

where ω_{SP} is known as the surface plasmon frequency. As $\omega \to \omega_{SP}$, the propagation constant $\beta \to \infty$ and the SPP wave becomes electrostatic in nature. This is because the group velocity $(v_g = \partial\omega/\partial\beta)$ approaches zero in this limit. In our analysis we have assumed both the permittivities to be real. However, the permittivity is in general a complex quantity as shown in Eq. (2.3). This leads to the propagation constant β being complex which results in damping of the SPP wave. Under this condition also, the expressions in this section are still valid and can be used to study the propagation of SPP waves.

In order to analyze the dependence of k_1, k_2 on ϵ_1, ϵ_2, we substitute Eq. (5.24) in Eq. (5.23) to get

$$\begin{aligned} k_1^2 &= -k_0^2 \left(\frac{\epsilon_{r1}^2}{\epsilon_{r1} + \epsilon_{r2}} \right) \\ k_2^2 &= -k_0^2 \left(\frac{\epsilon_{r2}^2}{\epsilon_{r1} + \epsilon_{r2}} \right) \end{aligned} \tag{5.26}$$

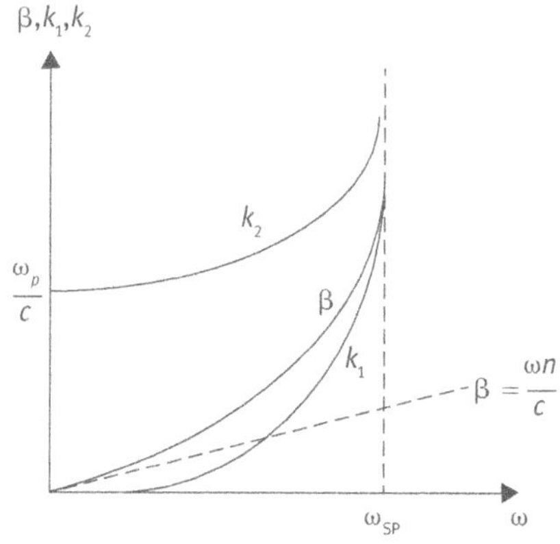

Fig. 5.2: The variation of β, k_1 and k_2 as a function of ω as given by Eqs. (5.24) and (5.26).

The condition $\epsilon_{r1} + \epsilon_{r2} < 0$ ensures that k_1, k_2 given by the above equation are always real as long as the permittivities are purely real. However, as noted above, the permittivities are in general complex which results in k_1, k_2 being complex and hence there is a lossy propagation of electromagnetic energy along both the x- and z-direction. Figure 5.2 shows how the quantities β, k_1, k_2 vary as a function of ω.

5.2 SPP on Multilayer Systems

Propagation of SPP waves on a single plane metal-dielectric interface is the simplest case of such a plasmon wave. There can be many more complex configurations including cylindrical interfaces. We will consider a simpler level of sophistication, namely a multilayer system consisting of two plane interfaces. There are two possibilities for such a configuration. As shown in the Figure 5.3, we can either have a metal layer sandwiched between two dielectric layers (IMI) or a dielectric layer sandwiched between two metal layers (MIM). Here, M stands for metal and I for insulator. Like in the case of the single interface SPP, here also the field components can be split into two modes: TM and TE. And again considering TM modes, the expressions for the field components in the three regions can be written as

$$H_y = \begin{cases} Ae^{i\beta x}e^{-k_3 z}, & z \geq a \\ \left(Ce^{k_1 z} + De^{-k_1 z}\right) e^{i\beta x}, & -a < z < a \\ Be^{i\beta x}e^{k_2 z}, & z \leq -a \end{cases} \tag{5.27}$$

where $k_1, k_2, k_3 > 0$ are the decay constants in the three regions and the z-dependence has been chosen to ensure that the wave decays as we move away from the two

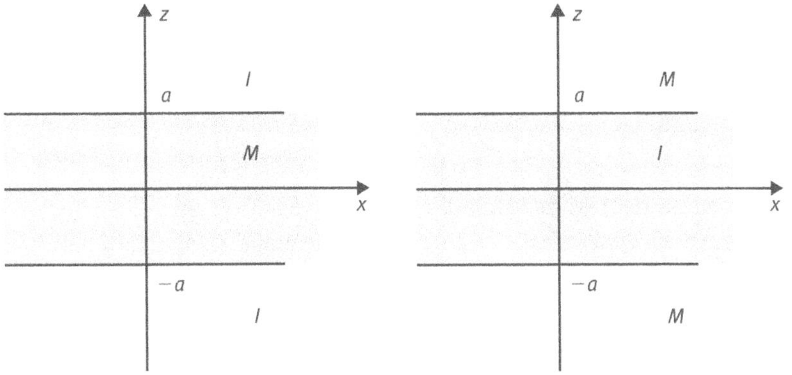

Fig. 5.3: The geometry of a multi-layer interface which supports surface plasmon waves.

interfaces at $z = \pm a$. Substituting this expression for H_y in Eq. (5.18), we get

$$E_x = \begin{cases} \left(iAk_3/\omega\epsilon_3\right) e^{i\beta x} e^{-k_3 z}, & z \geq a \\ \left(ik_1/\omega\epsilon_1\right)\left(-Ce^{-k_1 z} + De^{k_1 z}\right) e^{i\beta x}, & -a < z < a \\ -\left(iBk_2/\omega\epsilon_2\right) e^{i\beta x} e^{k_2 z}, & z \leq -a \end{cases} \tag{5.28}$$

and

$$E_z = \begin{cases} -\left(A\beta/\omega\epsilon_3\right) e^{i\beta x} e^{-k_3 z}, & z \geq a \\ -\left(Ce^{-k_1 z} + De^{k_1 z}\right)\left(\beta/\omega\epsilon_1\right) e^{i\beta x}, & -a < z < a \\ -\left(B\beta/\omega\epsilon_2\right) e^{i\beta x} e^{k_2 z}, & z \leq -a \end{cases} \tag{5.29}$$

Now we need to apply the same boundary conditions as we did for the TM mode for the single interface SPP wave described earlier. Equations (5.8) and (5.9) imply that E_x and H_y must be continuous across both the interfaces since we have assumed the free surface current to be zero. This gives four equations

$$\begin{aligned} Ae^{-k_3 a} &= Ce^{k_1 a} + De^{-k_1 a} \\ \frac{Ak_3}{\epsilon_3} e^{-k_3 a} &= -\frac{Ck_1}{\epsilon_1} e^{k_1 a} + \frac{Dk_1}{\epsilon_1} e^{-k_1 a} \\ Be^{-k_2 a} &= Ce^{-k_1 a} + De^{k_1 a} \\ -\frac{Bk_2}{\epsilon_2} e^{-k_2 a} &= -\frac{Ck_1}{\epsilon_1} e^{-k_1 a} + \frac{Dk_1}{\epsilon_1} e^{k_1 a} \end{aligned} \tag{5.30}$$

and since H_y must also satisfy the wave equation, Eq. (5.11), we get three additional equations

$$k_i^2 = \beta^2 - k_0^2 \epsilon_{ri} \tag{5.31}$$

where $i = 1, 2, 3$ and $k_0 = \omega\sqrt{\mu_0\epsilon_0}$. Equation (5.30) can be solved to obtain a relation between the k_is,

$$e^{-4k_1a} = \left(\frac{k_1/\epsilon_1 + k_2/\epsilon_2}{k_1/\epsilon_1 - k_2/\epsilon_2}\right)\left(\frac{k_1/\epsilon_1 + k_3/\epsilon_3}{k_1/\epsilon_1 - k_3/\epsilon_3}\right) \tag{5.32}$$

which coupled with Eq. (5.31) gives a unique solution for β for a given ω. Two interesting special cases are the MIM or IMI structure, where the materials II and III have the same permittivity. This implies $\epsilon_2 = \epsilon_3$ and $k_2 = k_3$. In this case, Eq. (5.32) splits into two equations

$$e^{-2k_1a} = \pm\left(\frac{k_1/\epsilon_1 + k_2/\epsilon_2}{k_1/\epsilon_1 - k_2/\epsilon_2}\right)$$

which can be rearranged to give

$$\tanh(k_1a) = -\frac{k_2\epsilon_1}{k_1\epsilon_2} \tag{5.33}$$

$$\tanh(k_1a) = -\frac{k_1\epsilon_2}{k_2\epsilon_1} \tag{5.34}$$

These two relations represent two different sub-modes of the TM mode of the SPP wave on such structures. In order to understand these sub-modes, let us reconsider Eq. (5.30). Dividing the 1st equation in this set by the 2nd and the 3rd by the 4th, we get the same equation

$$-\frac{\epsilon_2k_1}{k_2\epsilon_1} = \frac{Ce^{k_1a} + De^{-k_1a}}{Ce^{k_1a} - De^{-k_1a}} \tag{5.35}$$

Clearly, if $C = D$, this equation leads to Eq. (5.33) and if $C = -D$, we get (5.34). Now looking at Eq. (5.28) tells us that when $C = D$, then $E_x(z)$ is an odd function of z and when $C = -D$, then $E_x(z)$ is an even function of z. It is for this reason that the relation given by Eq. (5.33) is called the odd-mode and Eq. (5.34) is called the even-mode. Note that in the odd-mode, the fields H_y, E_z are even functions of z and vice-versa.

5.3 Excitation of SPP

The dispersion relation of an SPP wave on a single metal-dielectric interface is given by Eq. (5.24),

$$\beta = k_0\sqrt{\frac{\epsilon_{r1}\epsilon_{r2}}{\epsilon_{r1} + \epsilon_{r2}}} \tag{5.36}$$

where $\epsilon_{r1} > 0$ and $\epsilon_{r2} < 0$. If we also assume that $\epsilon_{r1} \geq 1$, then it certainly follows that $\epsilon_{r1}\epsilon_{r2}/(\epsilon_{r1} + \epsilon_{r1}) > 1$, which implies $\beta > k_0$. Now if we have a plane EM wave of wave number k_0 incident on the metal-dielectric interface from the dielectric side making an angle θ with the normal, the component of wave-vector parallel to the surface will be $k_0 \sin\theta$, which is definitely going to be smaller than β since $\sin\theta \le 1$ and $\beta > k_0$. Due to this reason, a plane EM wave cannot excite an SPP wave since for this to happen, the component of the wave vector parallel to the interface must be equal to β. Hence, in order to excite SPP waves, we must find a way to enhance the wave number of the incident EM wave. There are several ways to do this and we shall discuss a few of them.

While discussing Snell's laws and evanescent waves in Section 1.6, we found that when a plane EM wave undergoes total internal reflection on being incident on a rarer medium from a denser medium, the resultant EM wave in the rarer medium has an imaginary angle of refraction. As a result, the parallel component of the wave vector has a magnitude larger than the total wave number (see Eq. (1.41)). And as described above, this is precisely what we need in order to excite surface plasmons on a metal-dielectric interface. In order to use this technique, as shown in Figure 5.4a, a prism is placed above the metal (with a gap) on which we intend to excite SPP waves. Plane EM waves are made incident on the prism such that they undergo total internal reflection at its lower side and the resultant evanescent waves then get coupled to the metal surface in the form of SPP waves. For this to happen, the angle of the incident plane EM has to be finely tuned depending on the frequency of operation. This is because, as per Eq. (5.36), the β of the SPP wave is dependent on frequency and this β must match the k_x of the evanescent wave which depends on both frequency and angle of incidence (as shown in Eq. (1.41). This is known as the *Otto* configuration. Though this configuration is conceptually easy to understand, it has a small problem, which is that it is not easy to maintain the prism above the metal surface maintaining a uniform air gap. One situation where it is useful is when it is not desirable to have a direct contact between the prism and the metal, for example in situations where SPP waves are used to understand the properties of an unknown metallic substrate.

An alternative technique which is a little more complicated conceptually but much easier to use practically is known as the *Kretschmann* configuration and shown in Figure 5.4b. In this technique, a metal layer is deposited on to the lower surface of the prism itself. EM waves incident from the prism do not propagate into the metal layer since its permittivity is negative at the frequency of interest ($\omega \lesssim \omega_p$). Thus, they form a kind of an evanescent wave and the EM energy tunnels from the prism-metal interface to the other side of the metal surface. These magnitude of the parallel component of these tunneled waves is also larger than k_0 and hence can be matched with β under the right conditions.

This method of prism coupling can also be used for exciting SPP waves on multiple interface (both IMI and MIM structures). However, both these configurations of prism coupling have one common problem, which is that they have a significant loss of EM energy since only a small fraction of the incident EM waves get coupled to the metal-dielectric surface in the form of SPP waves. A large part of it is radiated back to the surroundings.

Another way to excite SPP waves is by using *dielectric gratings* on the metal or creating shallow gratings on the metal surface itself as shown in Figure 5.5. Using

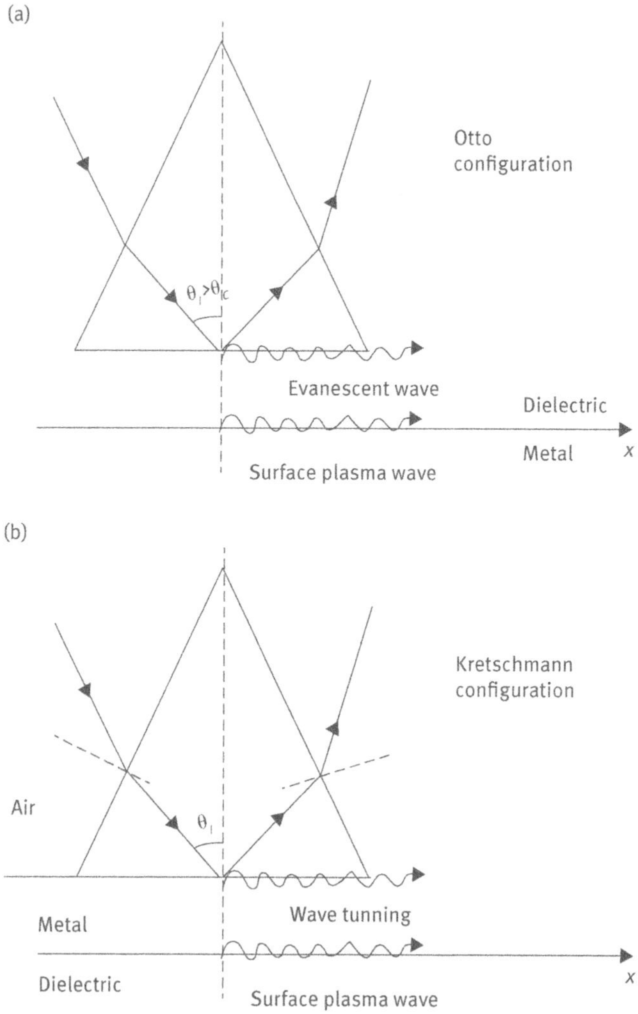

Fig. 5.4: The two ways in which a prism coupling can be used to excite surface plasmon waves.

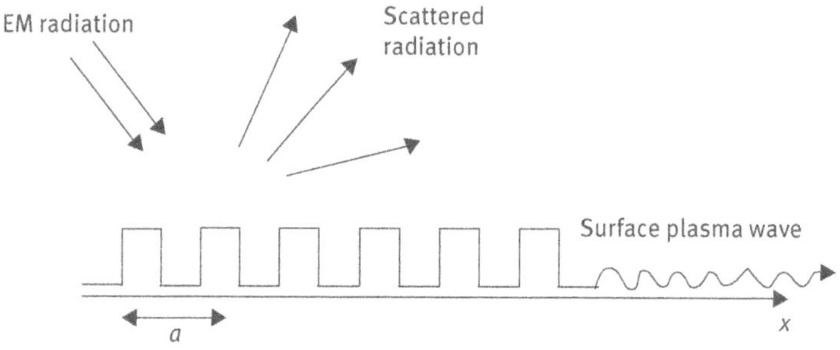

Fig. 5.5: How a diffraction grating can be used to excite surface plasmon waves.

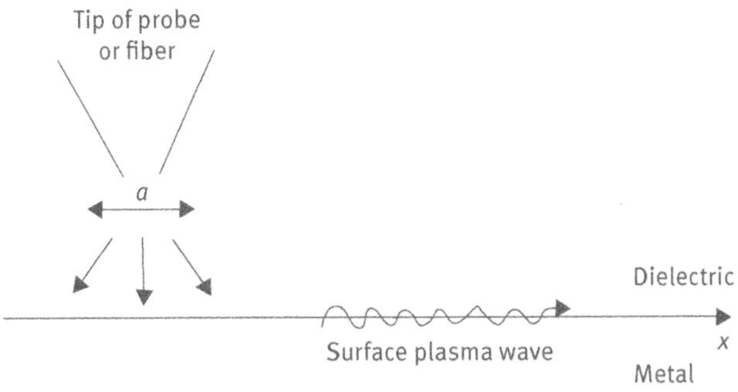

Fig. 5.6: How a probe tip can be used to excite surface plasmon waves.

Floquet theory [9], it can be shown that a grating with lattice constant a leads to creation of multiple components of the parallel wave number of the incident EM wave, $k_0 \sin\theta + 2\pi n/a$ [10], where $n \in \mathbb{Z}$. Now, the SPP wave is excited when the β given by Eq. (5.36) matches with one of these infinite wave numbers,

$$\beta = k_0 \sin\theta + \frac{2\pi n}{a} \tag{5.37}$$

Both these methods of SPP coupling (prisms and gratings) described above are extended in length. The prism surface and the gratings must be at least one wavelength long for EM coupling to take place. However, there is another method that can act as a point source for SPP coupling. As shown in the Figure 5.6, this method consists of a small probe tip of aperture size $a < \lambda_0$ on top of the metal-dielectric interface. Due to the small size of the tip, the near field EM waves emitted from this probe will have wave numbers larger than k_0 $(= 2\pi/\lambda_0)$ which can then couple to the SPP modes. This method is known as *near field excitation.*

5.4 Localized Surface Plasmon Resonance (LSPR)

In the previous section, we discussed the dispersion relation and other properties of surface plasmon polariton (SPP) waves propagating on a plane metal-dielectric interface. We saw that for such waves to exist, the permittivities of the metal and dielectric at the frequency of interest must be of opposite signs and this condition is achieved for frequencies just below the plasma frequency of the metal. However, excitation of such a SPP wave is tricky since its wave number is larger than the wave number of a plane EM wave. In order to circumvent this problem, we can use various techniques, namely prism coupling, grating coupling, near field excitation and a few more. We may now enquire about the nature of SPP waves on curved surfaces, but this situation is quite complex and very difficult to solve in general. A simpler limiting case is that of plane EM waves incident on a metallic spheres. Though this case is also very difficult to solve in general, we can find approximate solutions when the size of the sphere is much smaller compared to the wavelength of the incident EM radiation. Due to the curved surface of the sphere, the plasmon resonances can be excited by direct light illumination, in contrast to propagating SPPs, where the phase-matching techniques have to be employed.

In this section, we consider the case of a spherical particle of size $a \ll \lambda$ as shown in Figure 5.7a, and use Maxwell's equations to derive the expressions for the resulting surface plasmons. In this limiting case of $a \ll \lambda$, we can use the quasi-static approximation where the phase of the oscillating EM field is assumed to be practically constant over the entire particle volume. This leads to solutions known as the Localized Surface Plasmon Resonance (LSPR) and it is a non-propagating EM excitation of the conduction electrons of metallic nanostructures unlike the SPP waves on plane surfaces which are propagating. The curved surface of the particle exerts an effective restoring force on the driven electrons, so that a resonance can arise, leading to field amplification both inside and in the near-field zone outside the particle.

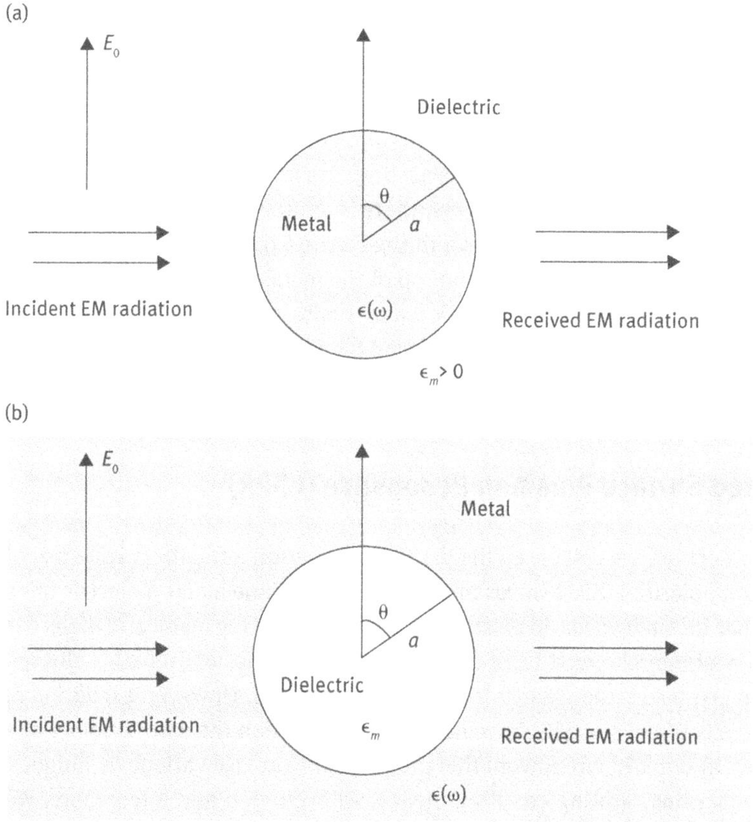

Fig. 5.7: Two configurations that support a localised surface plasmon resonance (LSPR). (a) A metal sphere of radius *a* embedded in a dielectric. (b) A dielectric hole of radius *a* embedded in a metallic region.

Consider a homogeneous, isotropic sphere of radius a and permittivity $\epsilon(\omega)$ located at the origin in a uniform, static electric field $\vec{E} = E_0\hat{z}$ and surrounded by a medium of permittivity ϵ_m. Since the electric field is assumed to be static in this approximation, the only equation we need to solve is the Laplace equation, $\nabla^2\Phi = 0$. If the size of the particles is comparable to the wavelength, then this approximation is not valid and one has to solve the full set of Maxwell's equations. This calculation is quite involved and is known as Mie scattering. In this section, we will not go into the details of Mie scattering and will only focus on the limiting case, $a \ll \lambda$.

Due to azimuthal symmetry, the general solution of the Laplace equation in spherical polar coordinates is given by (for details, see Appendix A)

$$\Phi(r,\theta) = \sum_{l=0}^{\infty}\left[A_l r^l + B_l r^{-l-1}\right] P_l(\cos\theta) \tag{5.38}$$

where $P_l(\cos\theta)$ are the Legendre Polynomials of order l, and A_l, B_l are constants to be determined using the boundary conditions. Since the fields must remain finite at the origin, the solutions for the potentials inside and outside the sphere can be written as

$$\Phi_{in}(r,\theta) = \sum_{l=0}^{\infty} A_l r^l P_l(\cos\theta)$$

$$\Phi_{out}(r,\theta) = \sum_{l=0}^{\infty}\left[B_l r^l + C_l r^{-l-1}\right] P_l(\cos\theta) \tag{5.39}$$

Now as $r \to \infty$, the effect of the sphere must die out to zero and what must remain is only the externally applied electric field, $\vec{E} = E_0\hat{z}$. Hence, as $r \to \infty$, we must have$\Phi_{out} \to -E_0 z = -E_0 r\cos\theta$. In Eq. (5.39), as $r \to \infty$, all the terms with coefficient C_l decay to zero and pose no problem in satisfying this condition. Comparing the terms with coefficient B_l, we get $B_1 = -E_0$ and $B_l = 0 \quad \forall l \neq 1$.

The potential Φ given by Eq. (5.39) must also satisfy the boundary conditions given by Eq. (1.33). Equality of the tangential components of $\vec{E}$ and the normal components of $\vec{D}$ demands that

$$-\frac{1}{a}\frac{\partial\Phi_{in}}{\partial\theta}\bigg|_{r=a} = -\frac{1}{a}\frac{\partial\Phi_{out}}{\partial\theta}\bigg|_{r=a} \tag{5.40}$$

$$-\epsilon\frac{\partial\Phi_{in}}{\partial r}\bigg|_{r=a} = -\epsilon_m\frac{\partial\Phi_{out}}{\partial r}\bigg|_{r=a} \tag{5.41}$$

The functions $P_l(x)$ form a complete orthonormal set of functions in the interval $x \in [-1,1]$ and hence, we can substitute Eq. (5.39) in the above boundary conditions and solve for the unknown coefficients (A_l, C_l) by comparing the coefficients of $P_l(x)$ term-by-term. From Eq. (5.40), we get

$$A_1 a = -E_0 a + C_1 a^{-2}$$

$$A_l a^l = C_l a^{-1-1} \qquad \forall l \neq 1 \tag{5.42}$$

and from Eq. (5.41), we get

$$\begin{aligned} \epsilon A_1 &= -\epsilon_m E_0 - 2\epsilon_m C_1 a^{-3} \\ \epsilon l A_l a^{l-1} &= -\epsilon_m (l+1) C_l a^{-l-2} \qquad \forall l \neq 1 \end{aligned} \tag{5.43}$$

where we have used the fact that $B_1 = -E_0$ and $B_l = 0 \quad \forall l \neq 1$. For $l \neq 1$, the above equations can be satisfied iff

$$\frac{1+l}{l} = -\frac{\epsilon}{\epsilon_m}$$

which is clearly not possible in general since ϵ/ϵ_m can be any real number whereas $1 + 1/l$ must be rational. This leads to the conclusion that

$$A_l = C_l = 0 \quad \forall l \neq 1.$$

Solving Eqs. (5.42) and (5.43) for A_1 and C_1, we get

$$\begin{aligned} \Phi_{in} &= -\frac{3\epsilon_m}{\epsilon + 2\epsilon_m} E_0 r \cos\theta \\ \Phi_{out} &= -E_0 r \cos\theta + \frac{\epsilon - \epsilon_m}{\epsilon + 2\epsilon_m} E_0 a^3 \frac{\cos\theta}{r^2} \end{aligned} \tag{5.44}$$

It can be seen that Φ_{out} represents the superposition of the applied field and that of a dipole located at the particle center.

The expression for Φ given by Eq. (5.44) gives finite results as long as $\epsilon + 2\epsilon_m \neq 0$. However, if $\epsilon + 2\epsilon_m$ does become zero at some frequencies, then the potentials blow up and we have what is known as the *Localized Surface Plasmon Resonance (LSPR)*. If we assume that ϵ is complex in general, the fields are resonantly enhanced when $|\epsilon + 2\epsilon_m|$ is minimum which happens when

$$\Re\left[\epsilon(\omega)\right] = -2\epsilon_m. \tag{5.45}$$

This condition is known as the *Frohlich condition* and the associated more is called the *dipole surface plasmon* of the metal nanoparticle. If the permittivity of the metal is given by Eq. (2.11),

$$\frac{\epsilon}{\epsilon_0} = 1 - \frac{\omega_p^2}{\omega^2}$$

this Frohlich condition is met when

$$\omega_{LSPR} = \frac{\omega_p}{\sqrt{1 + 2\epsilon_m/\epsilon_0}}. \tag{5.46}$$

Comparing this with the surface plasmon frequency, Eq. (5.25), we find that localized surface plasmon resonance is excited at a frequency within the frequency range for which planar SPP waves exist.

Here it is also interesting to consider the reverse condition: that of a dielectric hole in a metallic region as shown in the Figure 5.7b. All our analysis above holds good in this case too and requires only change. We interchange all instances of ϵ and ϵ_m. Thus, the Frohlich condition in this case becomes

$$\Re\left[\epsilon\left(\omega\right)\right] = -\frac{\epsilon_m}{2} \tag{5.47}$$

and the corresponding resonant frequency is given by

$$\omega_{LSPR,hollow} = \frac{\omega_p}{\sqrt{1+\epsilon_m/2\epsilon_0}} \tag{5.48}$$

which is clearly larger than the localized surface plasmon resonance frequency given by Eq. (5.46) and also outside the range of frequencies for which planar SPP wave exists.

5.5 Applications of Surface Plasmons

From Eq. (5.26), we can see that k_1 can be made larger than $k_0\sqrt{\epsilon_{r1}}$ by a suitable choice of ϵ_{r1} and ϵ_{r2}. The reciprocal of k_1 corresponds to the decay length of the SPP wave in the dielectric and the reciprocal of $k_0\sqrt{\epsilon_{r1}}$ is the wavelength of plane waves in the given dielectric. This implies that the SPP wave can confine electromagnetic energy to a region smaller than the wavelength corresponding to the operating frequency. This can have several important implications in the electronics industry where speed and miniaturization are key objectives. One specific application in this domain can be in computer chips [11] where transfer of information between various components of the circuit board can be a bottleneck. Conventional copper wires become very lossy at frequencies beyond the MHz range whereas information transfer is required at the THz range for faster communications. SPP waves can play a significant role here since they can support THz frequencies and also confine electromagnetic energy to sub-wavelength region. But here also, the currently available metals have a high imaginary component of permittivity at these frequencies and lot more research needs to be done to find materials with the desired properties.

Though applications of propagating surface plasmon waves are yet to find a practical implementation, several sensor devices have been developed using localized surface plasmon resonance (LSPR). As explained in Section 5.4, electromagnetic waves incident on a metallic spheres placed in a medium of permittivity ϵ_m is resonantly enhanced if its frequency satisfies the Frohlich condition given by Eq. (5.46). As shown in Figure 5.8, if we are transmitting plane waves from one end of the medium containing these metallic spheres and receiving them at the other end, we will see a dip in the received power at these resonant frequencies. Hence, if we transmit a broad range of frequencies, we can determine the permittivity of the medium using the frequency at which we see a dip in the received power. This is the basic working principle of the sensors based on LSPR. Several applications have been developed in this domain for detecting chemicals, biological entities like proteins and even individual molecules [12].

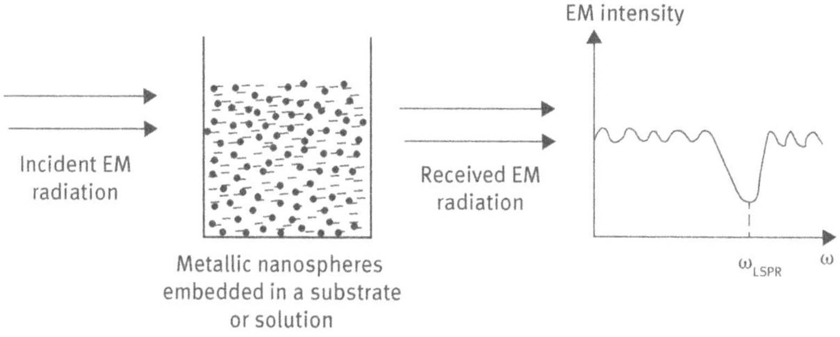

Fig. 5.8: This is a schematic of one of the ways by which LSPR is used in sensing applications.

Another very important application of plasmonics is in the domain of solar cells and this is again based on LSPR. A solar cell is an electrical device that uses the photovoltaic effect to convert visible light into electricity. The most common form of a solar cell is made using silicon but other materials have also been proposed over the years. In the photovoltaic effect, light is used to create free charge carriers within the material which then move through the material conducting electric current. This is different from the photoelectric effect where the free carriers are ejected out of the material. One of the major problems with solar cells is their low efficiency as most of the incident solar energy is lost without being converted to electricity. Plasmonics has the potential to significantly enhance this efficiency of energy conversion. It has been found that embedding metallic nanospheres in the photovoltaic material can significantly increase the absorption of solar energy at the resonant frequency and subsequent conversion into electricity [13].

Metallic nanospheres have also been found to help in enhancing the efficiency of infrared photography, which can be very helpful in penetrating foggy conditions and imaging dark areas. Traditional silicon based light sensors work at very high frequencies in the range of visible light. There are certain sensors which can directly capture infrared part of the spectrum, but these are very expensive. An affordable solution is based on using certain materials that can convert the incident infrared radiation to visible light which can then be imaged using the silicon based sensors. However, like in the case of solar cells, the efficiency of conversion of infrared to visible light is very low. But this can again be enhanced by using metallic nanospheres at the appropriate resonant frequency [14].

6 Spoof Surface Plasmons (SSP)

In the previous chapter, we studied the properties of surface plasmons on a planar metal-dielectric interface and found that for these waves to exist, the permittivity of the dielectric and metal must be of opposite signs. This condition is satisfied at frequencies slightly below the metal plasma frequency, since in this frequency range, the permittivity of the metal is given by

$$\frac{\epsilon}{\epsilon_0} = 1 - \frac{\omega_p^2}{\omega^2}$$

These surface plasmon waves are characterized by three quantities: β, k_1, k_2. β is the propagation constant along the interface, k_1 is the decay constant in the dielectric medium and k_2 is the decay constant in the metallic region. These quantities depend on the permittivities and frequency through these relations (Eqs. (5.24) and (5.26)),

$$\beta = k_0 \sqrt{\frac{\epsilon_{r1}\epsilon_{r2}}{\epsilon_{r1} + \epsilon_{r2}}}$$

$$k_1^2 = -k_0^2 \left(\frac{\epsilon_{r1}^2}{\epsilon_{r1} + \epsilon_{r2}} \right)$$

$$k_2^2 = -k_0^2 \left(\frac{\epsilon_{r2}^2}{\epsilon_{r1} + \epsilon_{r2}} \right)$$

where $k_0 = \omega\sqrt{\mu_0\epsilon_0}$ is the propagation constant in free space, $\omega < \omega_{SP} = \omega_p / \sqrt{1 + \epsilon_{r1}}$, $\epsilon_{r1} = \epsilon_1/\epsilon_0$ and $\epsilon_{r2} = \epsilon_2/\epsilon_0$. It is natural to enquire about the changes in the above quantities as ω becomes smaller and moves away from ω_{SP}. Clearly, as ω becomes smaller, ϵ becomes more and more negative leading to k_1 becoming smaller and k_2 becoming larger. When k_1 becomes small, it implies that the wave decays very slowly in the dielectric medium and when k_2 becomes large, it means that the wave decays very quickly in the metallic region. Thus, at very low frequencies ($\omega \ll \omega_p$), the SPP wave essentially becomes a plane EM wave at grazing incidence and looses its plasmonic properties. Thus, for all practical purposes, SPP waves exist only in a narrow band of frequencies near ω_p, which is in the optical and ultraviolet range for most metals. However, as it turns out, it is possible to mimic these surface plasmon waves at frequencies much lower than ω_p and it is these waves that are known as *Spoof Surface Plasmons* (SSP) [15, 16].

6.1 SSP at Low Frequencies

As mentioned above, the SPP waves theoretically exist for frequencies much below ω_{SP}[1] but become practically useless when $\omega \ll \omega_p$. Now, we also know from Eq. (2.10)

1 We must of course remember that at much lower frequencies, the metal permittivity is no longer given by Eq. (2.11) and here our analysis becomes invalid.

https://doi.org/10.1515/9783110570038-081

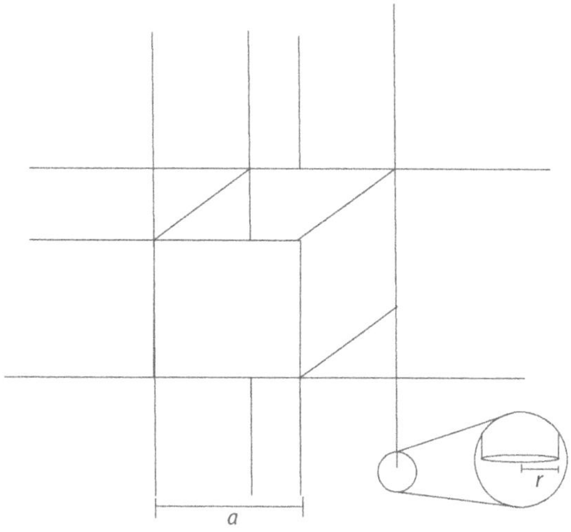

Fig. 6.1: 3D wire mesh used in creating materials with a negative effective permittivity.

that $\omega_p = \sqrt{Ne^2 f_0 / (m\epsilon_0)}$. Hence, the plasma frequency and range of practical validity of SPP waves depends on the density of electrons. If we could reduce N by some means, we could reduce ω_p and hence have practically meaningful surface plasmon waves at lower frequencies. A simple way to do this would be to drill holes in the metallic surface so that the *effective density* of electrons is reduced. Now of course, this effective medium idea is valid only when the wavelength of the EM wave is much larger than the dimensions of the holes so that the EM wave interacts with many holes at a time and hence, responds only to a averaged density. This is essentially what gave rise to the idea of mimicking surface plasmons at low frequencies (Spoof Surface Plasmons, SSP) by using metallic wire meshes [15] and structured surfaces [16].

One of the first method proposed to achieve an effective medium with ω_p in the GHz range was a 3D mesh of metallic wires [15] and is shown in Figure 6.1. The radius of each wire is r and the separation between adjacent wires is a. Thus, if the density of electrons in the wire is n, their effective density in the wire mesh structure as a whole is determined by the fraction of total volume occupied by the metal,

$$n_{\text{eff}} = n\frac{\pi r^2}{a^2}. \tag{6.1}$$

In addition to the effective density, we also need to estimate the effective mass of the electron. Due to the magnetic field of the wires, the electrons acquire an additional mass which is given by

$$m_{\text{eff}} = \frac{\mu_0 r^2 e^2 n}{2} \ln\left(a/r\right) \tag{6.2}$$

For aluminum wires of radius $r = 1\mu m$, spacing $a = 1mm$ and density $n = 1.806 \times 10^{29}\text{m}^{-3}$, we get

$$\begin{aligned} n_{\text{eff}} &= 5.674 \times 10^{23}\text{m}^{-3} \\ m_{\text{eff}} &= 2.01 \times 10^{-26}\ \text{kg} \\ &= 12.04 m_p \end{aligned}$$

which shows that the electron is now effectively more massive than a proton! Substituting this in the expression for the metal plasma frequency, we get

$$\omega_{p,\text{eff}} = 2.85 \times 10^{11}\ \text{rad}/\text{s}$$

which is much lower than $\omega_p = 2.39 \times 10^{16}$ rad/s for aluminum. A similar analysis can be done for the case of structured metallic surfaces [16]. The wavelength of plane waves corresponding to $\omega = 2.85 \times 10^{11}\ \text{rad}/\text{s}$ is $\lambda = 2\pi c/\omega = 6.614$ mm, which is about six times larger than the lattice constant $a = 1$ mm. The above analysis is valid for wavelengths much larger than the lattice constant and hence the surface wave modeled by this effective medium approach exists for frequencies much lower than $\omega_{p,\text{eff}}$.

In the case of SPP waves, we found that these waves exist only for frequencies below the plasma frequency. However, spoof surface plasmons can exist above the effective plasma frequency calculated using the effective density and effective electron mass given by Eqs. (6.1) and (6.2). Not only this, they have some very interesting properties at these higher frequencies which can play a very important role in future applications. Before we delve into these details, let us state two important properties of SSP waves that we will find do not hold for SPP waves at these higher frequencies,

1. SPP waves exist only in TM mode and there is no TE mode. This means that SPP waves have only an E_z component but no H_z.
2. The decay of SPP waves above and below the metal-dielectric interface can be accurately described by single exponentials. This means that the decay profile of the field components is given by $e^{k_1 z}$ and $e^{-k_2 z}$ below and above the interface respectively where $k_1, k_2 > 0$ are the decay constants.

These two properties approximately hold true for SSP waves at low frequencies, when the wavelength of the EM wave is much larger than the lattice constant. However, as the frequency becomes larger and the wavelength becomes comparable to the lattice constants, the discrepancies become important and can no longer be neglected [18, 19].

6.2 SSP at High Frequencies

In Section 5.1.2, we found that the SPP waves on both sides of a metal-dielectric interface can be described by single exponentials with decay constants $k_1, k_2 > 0$.

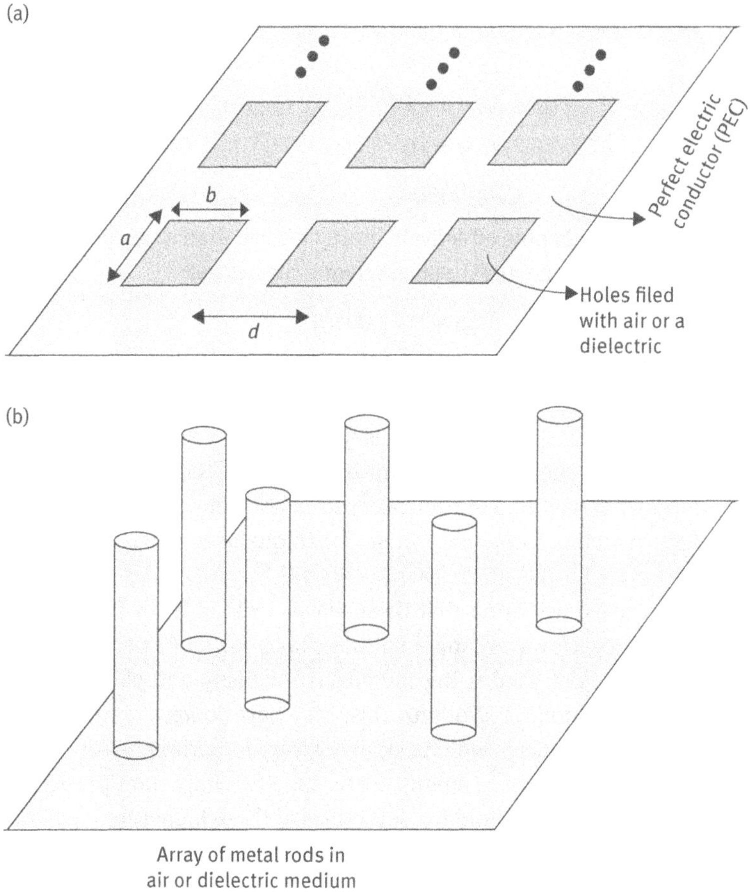

Fig. 6.2: Two different structured metallic surfaces which support spoof surface plasmons. (a) 2D array of dielectric holes in a PEC material. (b) 2D array of metallic rods in air or dielectric medium.

However, for this to be valid, it is important for the permittivities of the two sides have opposite signs. This is easy to achieve in the case of a metal-dielectric interface since the permittivity of a metal is negative at frequencies slightly below their plasma frequency. However, in the case of a spoof surface plasmon, we no longer have a planar metal-dielectric interface. We have either a wire mess or a structured metal surface as shown in Figure 6.2, where a large part of the interface has dielectric/air on both sides. Thus, the permittivity on both sides of the interface is the same for most of the points on the interface. Also, at the frequencies of interest in SSP, the metal permittivity is almost $-\infty$ and it behaves like a perfect electric conductor (PEC). Thus, on the parts of the interface containing the metal, we have a decay on the dielectric side and no field is present on the metal side. On parts of the interface containing dielectric on both

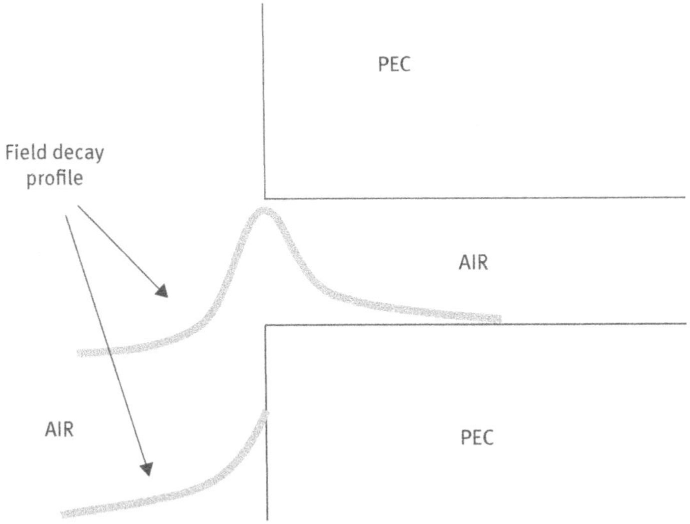

Fig. 6.3: The decay profile of SSP waves on a 2D array of holes in PEC material.

sides, we have a decay on both sides but this is not captured by single exponentials. One reason for this is that, as already mentioned, that for the decay to be captured by single exponentials, the permittivities on the two sides of the interface must be of opposite sign, which is clearly not possible if both sides have the same dielectric. Another reason is that single exponentials on the two sides of an interface imply a discontinuity (or at least a non-differentiability) of the field at the interface. This is again not possible if the two sides of the interface have the same dielectric medium.

Figure 6.3 gives a qualitative picture of the field decay profiles for the case of structured metallic surfaces. Here we specifically consider metal with infinitely deep holes. As can be seen in this figure, the actual decay profile of the wave on both sides of the interface is not given by simple exponentials and is very hard to be expressed analytically by a single function. But as it turns out, we can approximately express this decay profile as a superposition of several exponentials. Analytically estimating the decay constants of each of these exponential terms above the interface is quite challenging, but can be done so numerically. We know that the angular frequency and wave number of the electromagnetic wave in the dielectric medium must satisfy the dispersion relation $\omega = vk$, where v is the speed of light in this medium. Hence, if we take the values of the electric field vector on the interface at any point of time and takes it two-dimensional spatial Fourier transform, we will get the values of (k_x, k_y) for which the field components are dominant. Now, we know that $k_x^2 + k_y^2 + k_z^2 = \omega^2/v^2$ and using this we can find the value of k_z for each of the dominant values of (k_x, k_y). These values of $|k_z|$ are the decay constants of the wave above the interface and a superposition of these exponentials does in fact very well with the decay profile of the wave [18].

Finding the decay constants below the interface is much easier for this case and can be done analytically. If the holes in the metal are infinitely deep, they can essentially be modeled as infinitely long waveguides (see Section 1.8). For surface waves to exist on these interfaces, the frequency of the EM wave must be lower than the lowest cutoff of the waveguide modes. Otherwise most of the EM energy will go through these holes. For the case of square holes with side b, the value of k_z for various modes are given by

$$k_z = \sqrt{\frac{\omega^2}{v^2} - \left(\frac{m\pi}{b}\right)^2 - \left(\frac{p\pi}{b}\right)^2}$$

where $m, p = 0, 1, 2, \ldots$. It is not permissible for both m and p to be zero in these waveguides since then all the field components become zero. Thus, if $\omega < \pi v/b$, then k_z is imaginary and hence we have decaying components. This implies that the wavelength of the EM waves, $\lambda = 2\pi/k$, must be greater than twice the hole size, b. Again, taking a superposition of the leading terms gives us a decay profile that matches well with numerical simulations of these SSP modes [18]. One important result that comes from this analysis is that the SSP wave is a combination of both TM and TE modes unlike the SPP wave which was purely TM. Numerical simulations clearly show that the SSP wave has a significant magnitude of H_z which is another indicator for the presence of TE modes. Thus, the SSP wave is a hybrid mode.

The requirement stated above that for SSP waves to exist ($\lambda > 2b$) has another very interesting implication. SSP waves are essentially composed of EM waves whose components decay exponentially both above and below the interface. Now it is clear from above that the condition $\lambda > 2b$ ensures that the waves decay below the interface in the waveguide holes. But what about decay above the interface? Is this same condition sufficient to ensure this decay? It is important to note that the EM waves above the interface essentially consist of waves diffracted from the air holes. We have diffraction at every single hole on the interface and the SSP wave is a superposition of all these diffracted waves. So for the SSP waves to exist, these diffracted waves from each hole must consist of exponentially decaying components also known as evanescent waves. In order to address this issue, consider a plane at $z = 0$ on which the fields are known to zero everywhere except a small aperture whose coordinates are denoted by $\left(x', y', 0\right)$ and let the fields at any point in space be denoted by the function $u(x, y, z)$. According to the Rayleigh-Sommerfeld formulation [20], the field at any point in space can be written as a convolution between the field at the $z = 0$ plane and another function, $h(x, y, z)$,

$$u(x, y, z) = u(x, y, 0) * h(x, y, z) \tag{6.3}$$

$$h(x, y, z) = \frac{e^{ikr}}{2\pi r}\left(ik - \frac{1}{r}\right)\frac{z}{r} \tag{6.4}$$

where $r = \sqrt{(x - x')^2 + (y - y')^2 + z^2}$ is the distance of the given point in space (x, y, z) from the aperture located at $\left(x', y', 0\right)$. Taking a Fourier transform of Eq. (6.3), we get

$$U(k_x, k_y, z) = U(k_x, k_y, 0)\, H(k_x, k_y, z) \tag{6.5}$$

Now, we know that U must satisfy the Helmholtz equation given by

$$\frac{\partial^2 U}{\partial z^2} + \alpha^2 U = 0 \tag{6.6}$$

which can be easily solved to get

$$U(k_x, k_y, z) = A(k_x, k_y)\, e^{i\alpha z} + B(k_x, k_y)\, e^{-i\alpha z} \tag{6.7}$$

where $\alpha = \sqrt{k^2 - k_x^2 - k_y^2}$ and the functions A, B can be estimated using the known function $U(k_x, k_y, 0)$. Solutions of Eq. (6.6) are oscillatory or decaying in the z-direction depending on whether α^2 is positive or negative respectively. In order to rule out the possibility of non-physical growing solutions, we must have $B(k_x, k_y) = 0$ and $A(k_x, k_y) = U(k_x, k_y, 0)$. Comparing Eqs. (6.7) and (6.5), we get $H(k_x, k_y, z) = \exp(i\alpha z)$ and it can be shown that this is indeed the Fourier transform of $h(x, y, z)$ given by Eq. (6.4). Now, since we are interested in decaying solutions, we must have $k^2 < k_x^2 + k_y^2$. And from the discussion above, we know that in order to ensure decay of EM waves below the interface, we must have $\lambda > 2b \Rightarrow k < \pi/b$. So, both the condition for decay above the interface can be satisfied for $\lambda > 2b$ if the spatial frequencies k_x and k_y are roughly of the same order as $1/b$ or larger. The dominant spatial frequency components of the diffracted wave are related to the hole size through Fourier transform. A single square hole of size b is like a square pulse whose Fourier transform has dominant spatial frequency components below $1/b$. But if we have an array of holes with lattice constant, a, thats like a periodic square pulse of width, b, and periodicity, $a > b$. The Fourier transform of this periodic signal has dominant frequency components given by integer multiples of $1/a$. Hence, if we choose any $k < \pi/b$, it is likely to satisfy the requirement $k^2 < k_x^2 + k_y^2$ above the interface since most of the dominant values of k_x and k_y will be larger than $1/a$.

It is interesting to note that for a single pulse of width b, the major frequency components are below $1/b$. But when we have a periodic pulse train of period a, all these low frequency components superimpose to go to zero and only remaining components are integer multiples of $1/a$. So, although the diffracted waves from a single hole are radiating+evanescent but most of these radiating components surprisingly go to zero when we have a hole array. This is, in fact, what makes periodic structures and gratings so important in the area of photonics in general. There are several very interesting phenomenon that exist mainly because of periodicity in the material structure.

6.3 Self-collimation in SSP

As described in the previous section, the decay of SSP waves on both sides of the interface is not given by single exponentials but by a superposition of multiple components.

Fig. 6.4: Self-collimated beam of SSP waves on a 2D array of holes in a PEC material.
Source: Figure taken from [19] with permission.

Each of these components has a different (k_x, k_y) pair associated with it and the values of these pairs do not lie on any circle. Due to this, propagation of SSP waves is anisotropic in the sense that the wave does not propagate with the same intensity along all directions on the plane of the interface. This is unlike the case of SPP waves where there is no directionality and the flow of EM wave is isotropic. Presence of anisotropy in SSP waves is also intuitively meaningful. Since we have holes arranged on a square lattice, we clearly have at least two directions that have different structural properties (the sides of the square and the diagonal). And as it turns out, the primary flow of SSP waves happens along the diagonal with some energy also spreading out away from the diagonal. This anisotropy in SSP also has a very important feature known as *self-collimation* [17], which has no analogue in the case of SPP waves. What happens is that at a certain specific frequency, the SSP waves propagate in the same direction without spreading out (as shown in Figure 6.4). This is often known as diffraction-less propagation of EM waves. This phrase must however be treated with caution. The EM fields certainly get diffracted due to the holes present on the metal surface at all frequencies and it is this diffraction that is responsible for the existence and propagation of SSP waves. What is referred to by using the phrase *diffraction-less* here is merely a lack of spreading out of the EM energy along the surface. Interestingly, this self-collimation takes place at a frequency at which the corresponding wavelength is roughly twice the lattice constant of the square hole lattice. Thus, at this frequency, the effective medium approach described in Section 6.1 is not useful since its applicable only for wavelengths much larger than the lattice constant and in this limit the anisotropy of SSP is also largely absent.

For applications involving different kinds of SSP structures, it is important to be able to predict the self-collimation frequency. This can be done by using a method known as the *tight binding model* commonly used in the electronic band structure calculations in solid-state physics [22, 21, 20]. In solid-state physics, we have a two-dimensional square array of atomic orbitals which interact with each other through laws of quantum mechanics. And in the SSP case, we have diffracted electromagnetic waves from each square hole that interact with the diffracted waves from other holes through laws of electromagnetism. And as it turns out, in certain situations, quantum mechanics can bear a striking resemblance to electromagnetics. According to the tight binding model (under certain assumptions that we will not discuss in detail here and suggest the interested reader to see [20]), the eigenfrequencies of the SSP wave modes in the structure shown in Figure 6.2 are given by

$$\omega\left(\vec{k}\right) = \sum_{\vec{R}} e^{i\vec{k}\cdot\vec{R}} H_{mn}\left(\vec{R}\right) \tag{6.8}$$

where $\vec{k} = k_x\hat{x} + k_y\hat{y}$, is the wave vector in the xy-plane, $\vec{R}$ is a lattice vector and H_{mn} are the coupling parameters for the EM fields scattered from the mth and nth hole in the periodic structure. A lattice vector is the vector connecting the centers of different holes of the periodic lattice structure. In our periodic lattice shown in the Figure 6.2, the nearest neighbors are the ones that lie along the sides of the squares and hence the four lattice vectors connecting their centers are $\vec{R} = a\hat{x}$, $\vec{R} = -a\hat{x}$, $\vec{R} = a\hat{y}$ and $\vec{R} = -a\hat{y}$. Next in line are the holes along the diagonal and hence the four lattice vectors connecting them are $\vec{R} = a\hat{x} + a\hat{y}$, $\vec{R} = a\hat{x} - a\hat{y}$, $\vec{R} = -a\hat{x} + a\hat{y}$ and $\vec{R} = -a\hat{x} - a\hat{y}$. And similarly, we can write the lattice vectors corresponding to the holes farther and farther away. The more terms we keep in Eq. (6.8), the more accurate our expression for the eigenfrequencies will be. However, it is not practically possible to keep track of too many terms and for our purposes, it is sufficient to keep only the terms up to the second near neighbor. Due to symmetry, the hopping parameters, H_{mn}, corresponding to all the nearest neighbors are the same and so are they for the second nearest neighbors. Expanding Eq. (6.8) and keeping all these terms up to the second nearest neighbor interactions, we get

$$\begin{aligned}\omega(k_x, k_y) &\approx t_0 + t_1\left(e^{ik_xa} + e^{-ik_xa} + e^{ik_ya} + e^{-ik_ya}\right)\\ &\quad + t_2\left(e^{ik_xa}e^{ik_ya} + e^{-ik_xa}e^{ik_ya} + e^{ik_xa}e^{-ik_ya} + e^{-ik_xa}e^{-ik_ya}\right)\\ &= t_0 + 2t_1\left[\cos(k_xa) + \cos(k_ya)\right]\\ &\quad + 4t_2\cos(k_xa)\cos(k_ya)\end{aligned} \tag{6.9}$$

In order to use the above equation to find the eigenfrequencies corresponding to any value of the pair (k_x, k_y), all that we need are the values of the hopping parameters t_0, t_1, t_2. Since we have three unknown parameters, we need the values of ω for three pairs of (k_x, k_y). This part has to be done numerically as currently there are no known methods of analytically estimating these frequencies. For the sake of simplicity, we choose 3 commonly used high symmetry points where its fairly easy to get the values of ω through numerical simulations,

$$\begin{aligned}\Gamma &: k_x a = 0 = k_y a\\ X &: k_x a = \pi, \quad k_y a = 0\\ M &: k_x a = \pi = k_y a\end{aligned}$$

where Γ, M, X is the standard notation for these points in the domain of photonic crystals. Substituting the above in Eq. (6.9), we get

$$\begin{aligned}\omega_\Gamma &= t_0 + 2t_1 + 4t_2\\ \omega_X &= t_0 - 4t_2\\ \omega_M &= t_0 - 4t_1 + 4t_2\end{aligned} \tag{6.10}$$

Solving these three simultaneous equations, we get

$$\begin{aligned}t_0 &= -\frac{\omega_X}{2} - \frac{\omega_\Gamma + \omega_M}{4}\\ t_1 &= \frac{\omega_\Gamma - \omega_M}{8}\\ t_2 &= -\frac{\omega_X}{8} + \frac{\omega_\Gamma + \omega_M}{16}\end{aligned} \tag{6.11}$$

Our main purpose of carrying out this exercise was to find a method to predict the self-collimation frequency for SSP waves. We now have a dispersion relation for these waves (Eq. (6.9)) and all that we need to predict the self-collimation frequency is the corresponding value of (k_x, k_y). In order to get this, we need to plot the equifrequency contours corresponding to Eq. (6.9), as shown in Figure 6.5. The lines in this figure connect all the points in (k_x, k_y) space which correspond to the same eigenfrequency given by Eq. (6.9). So the main question is: which contour in this (k_x, k_y) space correspond to the self-collimation frequency?

As mentioned earlier, self-collimation is the frequency at which the SSP wave travels in the form of a nice beam without getting divergent. Hence, all the wave vectors, $\vec{k}$, corresponding to this frequency must travel in the same direction. But as we discussed in Section 1.7, all waves have two velocities: phase velocity and group velocity. Now for all these wave-vectors to have the same phase velocity $\left(\vec{v}_p = \omega\vec{k}/k^2\right)$, they must all point in the same direction which doesn't seem to hold for any equifrequency contour in Figure 6.5. And a single point in the (k_x, k_y) space cannot constitute a SSP wave for reasons described in Section 6.2. So do we have wave-vectors corresponding to a particular equifrequency contour all of whom have the same group velocity $\vec{v}_g = \partial\omega/\partial k_x \hat{x} + \partial\omega/\partial k_y \hat{y}$? At a certain point in the (k_x, k_y), the direction of phase velocity is given by the vector connecting that point with the origin. But, as explained in Section 1.7, the direction of group velocity is given by the direction normal to the equifrequency contour passing through that point. Hence, for a set of wave vectors

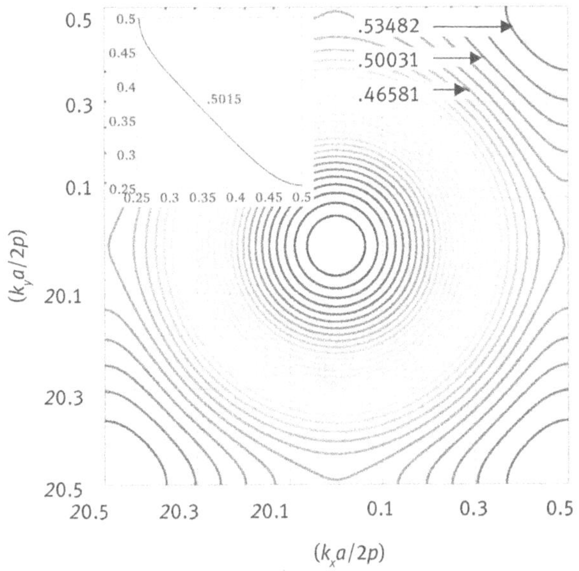

Fig. 6.5: Equifrequency contours of SSP waves on a 2D array of holes in a PEC material.
Source: Figure taken from [19] with permission.

to have the same group velocity, the corresponding equifrequency contour must be a straight line or very close to being straight. Such wave vectors indeed exist as shown in Figure 6.5.

So what we need to do now is use Eq. (6.9) to obtain an analytical estimate of the eigenfrequency for which the corresponding equifrequency contour is flat. And from the Figure 6.5, this condition is satisfied for the region where $d^2k_y/dk_y^2 = 0$ at $k_x = k_y$. Differentiating Eq. (6.9) with respect to k_x, we get

$$\begin{aligned} 0 &= 2t_1\left[\sin(k_x a) + \sin(k_y a)\frac{dk_y}{dk_x}\right] \\ &\quad + 4t_2\sin(k_x a)\cos(k_y a) + 4t_2\cos(k_x a)\sin(k_y a)\frac{dk_y}{dk_x} \end{aligned}$$

$$\Rightarrow \frac{dk_y}{dk_x} = -\frac{\sin(k_x a)\left[t_1 + 2t_2\cos(k_y a)\right]}{\sin(k_y a)\left[t_1 + 2t_2\cos(k_x a)\right]} \tag{6.12}$$

Substituting $k_x = k_y$ in the above equation, we get $dk_y/dk_x = -1$. Differentiating this equation once again in order to obtain an expression for d^2k_y/dk_x^2, we get

$$\begin{aligned} &\frac{d^2k_y}{dk_x^2} + \frac{\cos(k_y a)}{\sin(k_y a)}\left(\frac{dk_y}{dk_x}\right)^2 - \frac{4t_2\sin(k_x a)}{t_1 + 2t_2\cos(k_x a)}\frac{dk_y}{dk_x} \\ &\qquad + \frac{\cos(k_x a)\left[t_1 + 2t_2\cos(k_y a)\right]}{\sin(k_y a)\left[t_1 + 2t_2\cos(k_x a)\right]} = 0 \end{aligned} \tag{6.13}$$

Substituting $d^2k_y/dk_x^2 = 0$, $dk_y/dk_x = -1$ and $k_x = k_y$ in the above equation, we get

$$\cos(k_x a) = -\frac{2t_2}{t_1} \tag{6.14}$$

which implies that our required solution exists only if $|2t_2/t_1| \leq 1$. Inserting the above solution in Eq. (6.9) we get the required expression for the self-collimation (SC) frequency for SSP waves

$$\omega_{SC} = t_0 - 8t_2\left(1 - \frac{2t_2^2}{t_1^2}\right) \tag{6.15}$$

In the hole array shown in Figure 6.2, if $a = 10$mm and $b = 8.75$mm, the values of the eigenfrequencies at the three high symmetry points are approximately $\omega_\Gamma = 0$, $\omega_X = 13.47$ GHz and $\omega_M = 16.58$ GHz. For these values, the hopping parameters turn out to be $t_0 = 0.3625$, $t_1 = -0.0691$ and $t_2 = -0.0216$. Hence, for this structure, the self-collimation frequency is predicted to be at $\omega_{SC} = 15.05$ GHz which is very close to the numerically obtained value of 15.1 GHz [19].

7 Advanced Topics in Plasmonics

7.1 Negative Index Metamaterials (NIMs)

As the name suggests, NIMs are materials which have a negative index of refraction and were first proposed by V. G. Veselago in 1968 [24]. These materials are not naturally occurring but are artificially constructed using ideas similar to that used in the generation of spoof surface plasmon (SSP) waves discussed in Chapter 6. The refractive index of a material is given by $n = \sqrt{\epsilon_r \mu_r}$, where ϵ_r is the relative permittivity and μ_r is the relative permeability of the material. For naturally occurring materials, ϵ_r and μ_r are usually positive for most frequencies. However, as was discussed in Chapter 2, the permittivity of metals can become negative for some range of frequency near the visible spectrum. As shown in Chapter 6, we can mimic this negative permittivity behavior at much lower frequencies by drilling holes in the metal surface with an appropriate hole size and array size. In these cases, the permeability of the material is still positive, and hence the refractive index is imaginary. Due to this regular electromagnetic waves cannot travel inside the medium and all we have are surface plasmon waves. However, if we could make both the permittivity and permeability negative for the same frequency range, the product $\epsilon_r \mu_r$ will again be positive and hence the refractive index will be a real quantity. Such materials are also called *double negative metamaterials* (DNGs). In this case, we choose the negative sign for n to distinguish it from the usual case when ϵ_r and μ_r are positive. When the refractive index is real, we again have regular electromagnetic waves traveling through the medium but some of the properties of wave propagation in NIMs are very intriguing and very different from the case of Positive Index Materials (PIMs). NIMs have several applications, the primary among them being the superlens which can allow imaging beyond the diffraction limit. Other applications include metamaterial antennas which can help in miniaturization of transmitting devices, optical nanolithography and nanotechnology circuitry.

The simplest way to create a material with $\epsilon_r < 0$ is to have a wire mesh of appropriate dimensions (as shown in Figure 6.1). For $\mu_r < 0$, we need to have an array of split-ring resonators (SRRs). The wire mesh creates an effective negative permittivity by coupling to the electric field component of the electromagnetic wave as described in Chapter 6. Similarly, an array of split-ring resonators (shown in Figure 7.1) creates an effective medium with negative permeability by coupling to the magnetic field component of the electromagnetic wave. These SRRs are designed so as to mimic the magnetic response of atoms in natural materials. The magnetic field of an electromagnetic wave induces rotating currents in the SRR coils and a coupling of the currents between various SRRs leads to the effective medium behavior which can be suitably tuned to lead to negative permeability. The rings of the SRRs are split and not closed so that they can support wavelengths much larger than their own diameter.

https://doi.org/10.1515/9783110570038-093

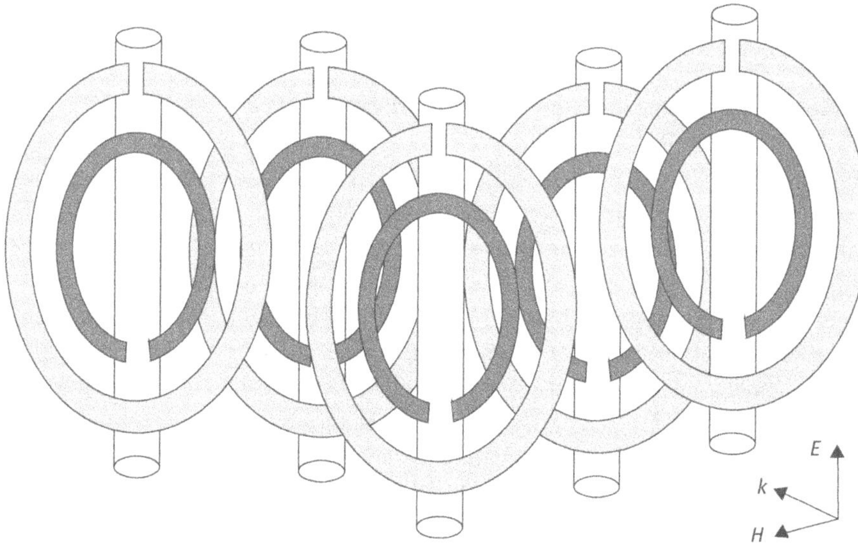

Fig. 7.1: This figure shows a 2D array of split ring resonators used in creating a material with negative effective permeability. This array coupled with the 3D wire mesh shown in Figure 6.1 can be used in creating negative index materials (NIMs).

One of the limitations of NIMs is the size of the components used which determines the wavelength/frequency for which the composite material behaves like an NIM. As was discussed in Chapter 6, these components have to be much smaller than the wavelength at which they have negative permittivity and permeability. And hence, it becomes increasingly difficult to support this behavior at optical frequencies since the corresponding wavelengths are very small. It is very easy to create an NIM material that operates in the microwave region because then we can use the regularly available components in any electronics laboratory. But at infrared and optical frequencies, one has take recourse to lithographic techniques, nanorods and nanoparticles which are not easy to manipulate.

In order to understand the electromagnetic properties of materials with negative index of refraction, let us consider two of the Maxwell's equations in free space

$$\vec{\nabla} \times \vec{E} = -\mu \frac{\partial \vec{H}}{\partial t}$$
$$\vec{\nabla} \times \vec{H} = \epsilon \frac{\partial \vec{E}}{\partial t}$$

Taking a Fourier transform in both the time and spatial domain, we get

$$\vec{k} \times \vec{E} = \omega\mu\vec{H}$$
$$\vec{k} \times \vec{H} = -\omega\epsilon\vec{E} \tag{7.1}$$

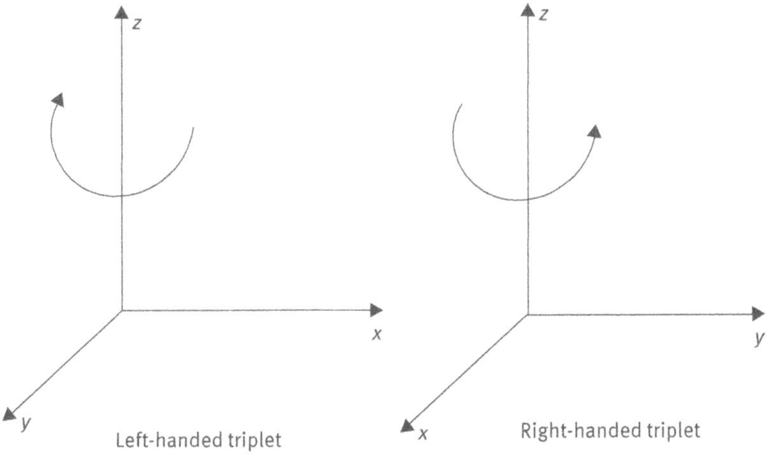

Fig. 7.2: This figure shows the difference between left-handed and right-handed triplets.

From the above equations, we can see that $\vec{E}$, $\vec{H}$ and $\vec{k}$ form a right-handed triplet of vectors if $\epsilon, \mu > 0$. But if both $\epsilon, \mu < 0$, then these vectors form a left-handed triplet (as shown in Figure 7.2). The vector $\vec{k}$ determines the direction of phase velocity of the electromagnetic wave as it propagates. An electromagnetic wave also carries momentum/energy and its direction of flow is governed by the Poynting vector given by $\vec{S} = \vec{E} \times \vec{H}$, whose direction is opposite to that of $\vec{k}$ for NIMs. Thus, it can be seen that for NIMs which are left-handed materials, the direction of flow of electromagnetic energy (group velocity) is opposite to the direction of phase velocity. What this means is that when the E-field is along the $+x$ direction and the H-field is along the $+y$ direction, the flow of electromagnetic energy will be along $+z$ direction for both PIMs and NIMs. However, the phase of the wave will move along $+z$ direction for PIMs and along $-z$ direction for NIMs. Due to this property, NIMs are sometimes referred to as *backward-media* or materials with *negative group velocity*. This can have very important implications during refraction of electromagnetic waves when they are incident from a PIM to an NIM. Here it is important to note that the wave equation remains unchanged as long as the product of permittivity and permeability is positive, $\epsilon\mu > 0$,

$$\nabla^2 \vec{E} = \epsilon\mu \frac{\partial^2 \vec{E}}{\partial t^2} \tag{7.2}$$

When a plane electromagnetic wave is incident from one medium to another, the angle of incidence and refraction (or transmission) are connected by what is known as Snell's law,

$$n_1 \sin\theta_I = n_2 \sin\theta_T \tag{7.3}$$

where n_1, n_2 are the refractive indices of the two media, $-\pi/2 < \theta_I < \pi/2$ is the angle of incidence and $-\pi/2 < \theta_T < \pi/2$ is the angle of transmission. When $n_1 > 0$ (PIM)

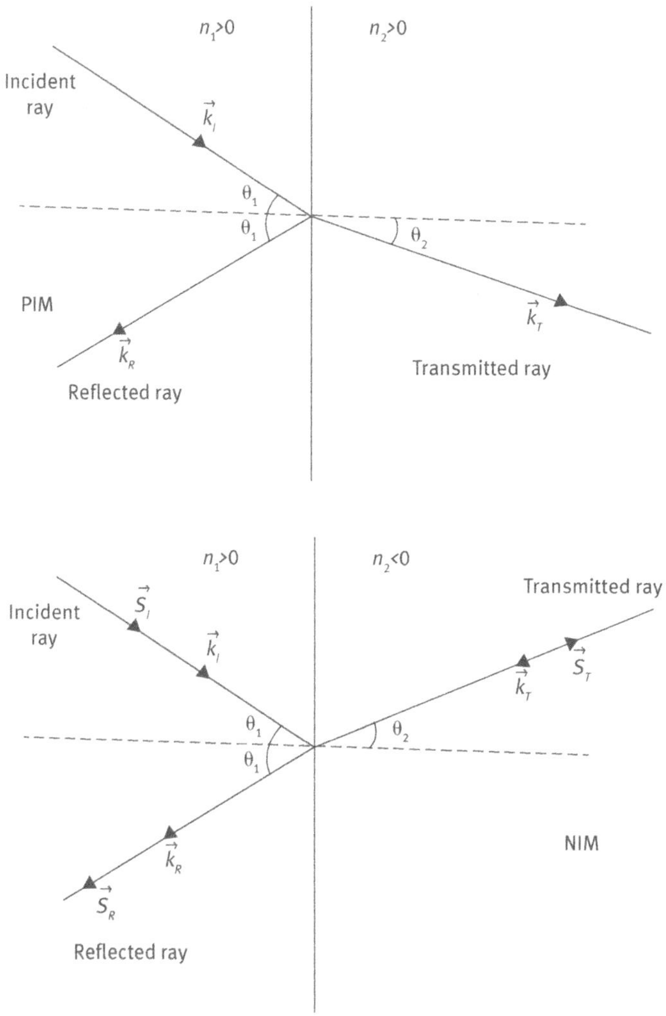

Fig. 7.3: This figure shows the relation between incident, reflected and transmitted waves in a negative index material (NIM).

and $n_2 < 0$ (NIM), we can clearly see that θ_I and θ_T cannot be of the same sign and the incident and transmitted wave happen to be on the same side of the normal as shown in Figure 7.3. And if $\epsilon_2 = -\epsilon_1$ and $\mu_2 = -\mu_1$, there is no reflected wave and the entire electromagnetic wave is transmitted from one medium to another but with a change of direction.

A very special property of NIMs called *superlens* emerges when we have $n_2 = -n_1$. This terminology is a little misleading since a lens is usually supposed to focus parallel light rays coming from infinity. This NIM superlens instead focuses a point source of

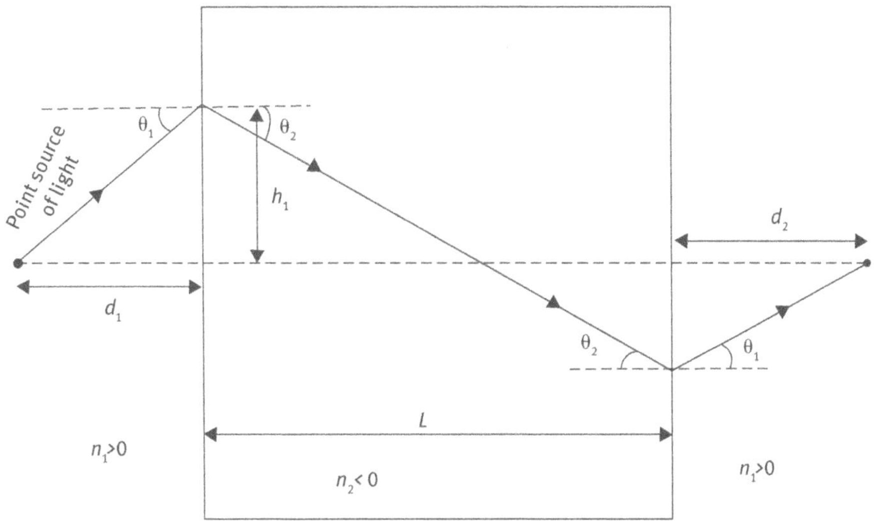

Fig. 7.4: This figure shows an NIM slab being used as a superlens.

light on one side to its other side into another point. As shown in the Figure 7.4, we consider a point source of light (or a point object that reflects light incident on it) placed in a material of refractive index $n_1 > 0$ at a distance d_1 from a NIM slab of thickness L and refractive index $n_2 < 0$. As noted above and as shown in the Figure 7.4, light incident from the point source onto the NIM slab will be refracted in a direction thats on the same side of the normal as the incident wave and the light ray is again focussed on the other side of the slab at a distance d_2. Using simple rules of trigonometry, we get

$$d_2 = (L - d_1)\,\frac{\tan\theta_1}{\tan\theta_2} \tag{7.4}$$

Now, if we want the point source of light to be focussed on the other side of the NIM slab, all rays emanating from this point source must meet at the same point on the other side. This means that d_2 must be independent of the angle of incidence. From Eq. (7.4), we can see that this happens only when $\theta_1 = \theta_2$, which in turn implies that $n_2 = -n_1$. Here, it is important to note that $\theta_I = \theta_1$ and $\theta_T = -\theta_2$. If $n_2 \neq -n_1$, then paraxial rays coming from the point source can still be focussed at a point on the other side of the NIM slab. For paraxial rays, we have $\tan\theta_1 \approx \theta_1$ and $\tan\theta_2 \approx \theta_2$. From Snell's law, we also know that

$$\frac{\theta_1}{\theta_2} \approx \frac{\sin\theta_1}{\sin\theta_2} = -\frac{n_2}{n_1}$$

Thus, the distance from the NIM slab at which the paraxial rays are focussed is given by

$$d_2^{\|} = -(L - d_1)\,\frac{n_2}{n_1} \tag{7.5}$$

This superlens property of NIMs which enables it to focus a point source into another point image can be very useful in imaging beyond the diffraction limit. In conventional cameras, when light is incident from two points that are closely spaced, the electromagnetic wave from these two points undergoes diffraction at the camera aperture which makes it difficult to increase the camera resolution beyond a certain limit. However, since NIM materials have the property of negative refraction, they can refocus two closely space points into another set of closely spaced points on the image screen thereby defying the diffraction limit. This can be particularly useful in the imaging of biomolecules like DNA, proteins, etc and even viewing of live biological cells. This lens made of an NIM slab is also sometimes called a *perfect lens* since it can focus not only propagating waves but also near field evanescent waves, which can sometimes contain very useful information. Though this NIM lens has very interesting and useful properties, realizing this in practice is very challenging and has not been successful so far due to two reasons. Firstly, most NIM materials are lossy and dispersive due to which the actual material properties are very different from what has been assumed in the above analysis. Secondly, the required condition of $n_2 = -n_1$ is very stringent and even a slight deviation leads to an image that is highly degraded.

The fact that NIMs are highly dispersive can be understood by analyzing the density of electromagnetic fields in such materials. In non-dispersive media, the time-averaged energy density of electromagnetic fields is given by

$$U_{nd} = \frac{1}{4}\left[\epsilon\left|\vec{E}\right|^2 + \mu\left|\vec{H}\right|^2\right] \tag{7.6}$$

which would lead to a negative energy density for NIMs since in these materials, $\epsilon, \mu < 0$. However, since energy density must be positive by convention, we take recourse to the expression for energy density for a quasi-monochromatic wavepacket in a dispersive medium which is given by

$$U_d = \frac{1}{4}\left[\frac{\partial(\omega\epsilon)}{\partial\omega}\left|\vec{E}\right|^2 + \frac{\partial(\omega\mu)}{\partial\omega}\left|\vec{H}\right|^2\right] \tag{7.7}$$

where the derivatives are evaluated at the central frequency of the wavepacket. Now, for NIMs to have positive energy density, we must have

$$\frac{\partial(\omega\epsilon)}{\partial\omega} > 0 \qquad \text{and} \qquad \frac{\partial(\omega\mu)}{\partial\omega} > 0$$

which implies that $\partial\epsilon/\partial\omega > |\epsilon|/\omega$ and $\partial\mu/\partial\omega > |\mu|/\omega$. Therefore, NIMs must be highly dispersive.

7.2 Surface-Enhanced Raman Scattering (SERS)

When electromagnetic energy, in the form of photons, is incident on an atom or molecule, it can either be resonantly absorbed or scattered. Resonant absorption takes place when the energy of the photon is equal to the difference between two energy levels of the atom. The atom then stays in the excited state for a certain period of time and the photon is re-emitted with the atom coming back to its ground state. This phenomenon is known as *fluorescence*. When the photon energy is not equal to the difference between two energy levels of the atom, the photon undergoes a non-resonant process known as *scattering* which can either be elastic or inelastic. In the case of elastic scattering, the energy of the emitted photon is same as the energy of the incident photon. This elastic scattering is the usual process that we experience in our day to day life in the form of reflection and refraction from various kinds of surfaces. When elastic scattering takes place through interaction with particles which have a size much smaller than the wavelength of incident electromagnetic waves, it is called *Rayleigh scattering*. And when the particle size is comparable to the wavelength, it is called *Mie scattering*. When the particle size is much larger than the wavelength of light, the scattering process can be well captured by the simple rules of geometrical optics. The phenomenon of non-resonant scattering where the energy of emitted photon is different from that of incident photon is known as *Raman effect*. This phenomenon was experimentally demonstrated by C. V. Raman for which he was awarded the Nobel Physics Prize in 1930. Theoretical prediction of the existence of inelastic scattering was given by Adolf Smekal in 1923.

The fraction of photons which undergo inelastic scattering on being incident on a atom/molecule is incredibly small, roughly about 1 in a million. Inelastic scattering usually occurs when the photons couple to the vibrational states of an atom/molecule, which typically have an energy in the infrared region of the electromagnetic spectrum. So when photons in the near-infrared or visible frequency region are incident on atoms/molecules, some of them couple to the vibrational states, get absorbed and re-emitted at a higher or lower frequency/energy. If the released photon has a lower energy than the incident photon, the process is called Stokes scattering. And when the released photon has a higher energy than the incident photon, the process is called anti-Stokes scattering. At thermal equilibrium, most atoms/molecules are more likely to be found in their ground state and hence it is more likely for the photons to be re-emitted at a lower energy. Due to this reason, the peaks corresponding to the Stokes scattering are usually stronger than the peaks corresponding to the anti-Stokes scattering. The Raman effect is called non-resonant scattering since it does not require the incident photons to have a particular frequency equal to the difference between the atomic energy levels.

In a typical experimental setup, a laser beam of suitable frequency is made incident on the sample to be analyzed. The scattered electromagnetic waves are made to pass through a lens, filtered through a suitable filter (so as to remove the electromagnetic energy at the incident frequency) and the remaining scattered photons (called *Raman spectrum*) are collected to be further analyzed. Each substance has a unique Raman spectrum depending on its chemical composition and vibrational state. Raman spectroscopy has found immense applications in many domains of science and engineering mainly for the purpose of analysis of various substances. It is very useful in the non-destructive analysis of paintings and other works of art by helping in analyzing the various pigments used. It can also be used in the study of crystal structures by helping in analyzing the orientation of crystals.

One of the limitations of Raman spectroscopy is that by itself it is a weak effect. As noted above, only about 1 in a million incident photons undergo inelastic scattering. Several techniques have been proposed to enhance this process and one of the them is known as Surface Enhanced Raman Scattering (SERS). In this technique, the scattering process can be significantly enhanced by adsorbing the molecules to rough metallic surfaces or plasmonic nanostructures. In some cases, the inelastic scattering can be enhanced by a factor as high as 10^{10}, thereby making it possible to detect single molecules. The exact mechanism of SERS is still a matter of debate in the scientific community, but it is widely believed that localized surface plasmons (LSPR, discussed in Section (5.4)) have an important role to play in this process. It is also possible that the enhancement is because the molecules form chemical bonds with the surface, but this cannot explain enhancement in cases where no such chemical bonds are possible.

In the case of LSPR, what happens is that the electromagnetic waves incident on a collection of nanoparticles get resonantly enhanced if their frequency satisfied the Frohlich condition given by Eq. (5.45). Now if we place the substrate to be analyzed in this region of enhanced electromagnetic field, it is likely that the substrate will absorb many more photons than what it would have done in free space. Similarly, the photons emitted as a result of inelastic scattering can also be enhanced by a similar process if their frequency is again close to incident frequency. Thus, the LSPR effect works by enhancing both the incident electromagnetic energy as well as the emitted photons. However, for this process to work, the frequencies of the incident and emitted photons must satisfy the Frohlich condition at least approximately. Though the SERS effect works in a very narrow range of frequencies, it is still very useful in detecting biomolecules and other chemicals. For these experiments, typically gold or silver nanorods or nanoparticles are used since their resonant frequencies lie in the visible and near-infrared range, thereby providing a very large amount of enhancement.

7.3 Particle Traps

Confinement of charged particles is a very important problem from the perspective of both fundamental physics as well as wide ranging applications which include mass spectroscopy, quantum computing and DNA sequencing. There are several mechanisms for trapping charged particles, but Paul trap is perhaps the most important one used in the applications mentioned. Earnshaw's theorem prevents static electric fields from trapping charged particles in 3D space by showing that a static electric potential only has saddle points and no local minima/maxima. As shown

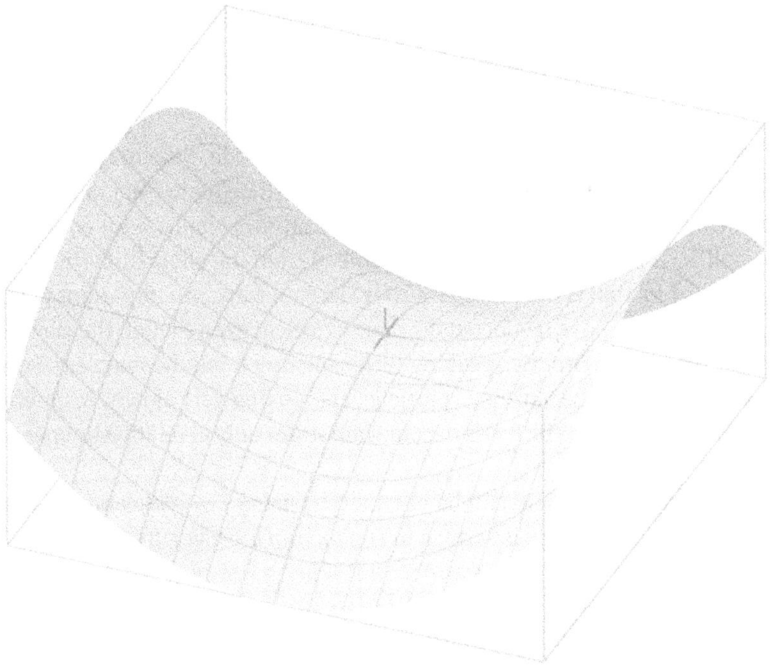

Fig. 7.5: Saddle point.

in Figure 7.5, at a saddle point, the particle may be confined along one direction but slips out of the device along the other direction.

The main idea behind Earnshaw's theorem is easy to see. Consider a hypothetical static force field, $\vec{E}$, which is able to confine a particle with positive charge, Q, at a particular point in 3D space. This means that in a neighbourhood around this point, the force field must either be pointing inwards or be spiralling inwards towards this point (see Chapter 8 for more details). Thus, either the divergence of $\vec{E}$ at this point must be negative or the curl of $\vec{E}$ must be non-zero. However, for the case of static fields, we know from Gauss' law and Faraday's law that in a charge free region, the gradient as well as the curl of the electric field is zero i. e., $\vec{\nabla} \cdot \vec{E} = 0 = \vec{\nabla} \times \vec{E}$. Thus, in the electrostatic case, all points of a charge free space must be saddle points.

A Paul trap [25] is a simple resolution to this problem and confines charged particles through the use of spatially linear time-periodic electric fields. Wolfgang Paul was awarded the Physics Nobel Prize in 1989 along with Hans G. Dehmelt for development of this novel idea. There is an alternative trap known as Penning trap which uses a combination of static electric and magnetic fields, but it is not very widely used since generation of magnetic fields requires bulky current coils which are difficult to miniaturise and maintain. A conventional Paul trap does not use magnetic field but variations have been proposed over the past few years for the purpose of trapping anti-hydrogen.

In a Paul trap, the externally applied potential is of the form

$$\phi(x, y, z, t) = \frac{U_0 + V_0 \cos \omega t}{r_0^2 + 2z_0^2} \left(x^2 + y^2 - 2z^2\right) \tag{7.8}$$

where U_0 is the static electric potential, V_0 is the time-periodic potential, r_0 is the radial extent of the trap, z_0 is the axial extent and ω is the RF frequency (usually in the MHz range). Of course, the potential given by Eq. (7.8) satisfies the Laplace equation, $\nabla^2 \phi = 0$. Thus, if $V_0 = 0$, this potential is incapable of trapping charged particles. It is the introduction of V_0 that is the primary novelty in Paul traps. An important point to note over here is that we have neglected the magnetic field even though the electric potential is time-dependent. Strictly speaking this is incorrect, but the magnetic field effects are negligible mainly because our particles are moving very slowly. Magnetic field effects of time-varying electric fields become important only at relativistic speeds of the test particle.

The corresponding electric field can be obtained by using $\vec{E} = -\vec{\nabla}\phi$ and thereby, the equations of motion of a charged particle in such a trap are given by

$$\begin{aligned} \ddot{x} &= [-p + 2q \cos \omega t]\, x \\ \ddot{y} &= [-p + 2q \cos \omega t]\, y \\ \ddot{z} &= [2p - 4q \cos \omega t]\, z \end{aligned} \tag{7.9}$$

$p = 2QU_0 / \left[M\left(r_0^2 + 2z_0^2\right)\right]$ and $q = -2QV_0 / \left[M\left(r_0^2 + 2z_0^2\right)\right]$ are normalized constants and M, Q are the mass and charge of the trapped particle. The three equations in Eq. (7.9) have the same form known as the Mathieu's equations [4,6], which also find immense applications in solving for the quantum wave function of electrons in semiconductor devices and the electromagnetic band gaps in photonic crystals. The solutions of Eq. (7.9) can be either bounded (periodic or aperiodic) or unbounded (exponentially growing with time) depending on the values of p, q and ω. Thus, it is very important to carefully choose these parameters. Though the presence of unboundedness can be a nuisance, it is also very important in mass spectroscopy. If we have a collection of different charged particles in a trap, by carefully tuning these parameters, we can selectively eject charged particles from the trap and thereby identify them since the parameter at which a charged particle gets ejected gives a good estimate of its M/Q value.

As shown in Eq. (7.9), though the Paul trap is a 3D device, the equations of motion of charged particles get decoupled along the three coordinate axes and can be solved separately. The equation of motion along the x-direction is given by

$$\frac{d^2x}{dt^2} + [p - 2q \cos \omega t]\, x = 0 \tag{7.10}$$

For a fixed value of ω, Eq. (7.10) does not have bounded solutions for all p, q. Also, the bounded solutions of Eq. (7.10) are, in general, aperiodic. However, for any given q, there is a countable set of values of p denoted by $a_0(q), a_1(q), a_2(q), \ldots$ for which Eq. (7.10) has periodic solutions which are even with respect to time, i. e., $x(-t) = x(t)$. Similarly, there is another countable set of values of p denoted by $b_1(q), b_2(q), b_3(q), \ldots$ for which the equation has periodic solutions which are odd, i. e., $x(-t) = -x(t)$. These two countably infinite sets of values $(a_0, a_1, a_2, \ldots)$ and $(b_1, b_2, b_3, \ldots)$ are called the characteristic values of the Mathieu equation. For $q > 0$, these values form a nice ordered set and we have, $a_0 < b_1 < a_1 < b_2 < \ldots$. Importantly, for values of $p \in (a_r, b_{r+1})$, where $r = 0, 1, 2, \ldots$, the solutions to Eq. (7.10) are aperiodic and bounded. And for other values of p, i. e., $p \in (b_r, a_{r+1})$ for $r = 1, 2, \ldots$, the solutions grow exponentially to infinity. Thus, on the $p - q$ space, the set of values for which the solutions are periodic have a measure zero.

For $p = 0$, the solutions are stable if $0 \leq |q| \leq q_c$ where $q_c \approx 0.225\omega^2$. In a Paul trap, this corresponds to the case when the DC field is zero and the particles experience only the RF field. We can see from Eq. (7.9) that the sign of both the DC and AC terms is opposite for the force equation in x and z directions. If we consider these equations to be equivalent to Eq. (7.10), we will get $p > 0$ in one case and $p < 0$ in another case, and similarly for q. For small values of p and q, the DC term can be stabilizing or destabilizing depending on its sign, but the AC term is always stabilizing (as we know from the ponderomotive theory described in Section 3.4). If $p < 0$ in Eq. (7.10), since the DC field alone would lead to unbounded solutions, any arbitrarily small positive value of q will not be sufficient to confine the particles and we need something strong enough to counter the effect of the DC field. But obviously, q cannot also be arbitrarily large since it must be between the two characteristic values. Thus, when $p < 0$, there is a non-zero positive lower bound and an upper bound on the value of q for the solutions of Eq. (7.10) to be stable. For $p > 0$, there is no lower bound on $|q|$ since both the DC and the RF field contribute towards confining the particles. However, this is true only for 1D motion and restrictions might arise when we consider all the three directions simultaneously.

Equation (7.10) belongs to the general class of linear ordinary differential equations with periodic co-efficients and thus can be solved by using the Floquet theorem. According to this theorem, the solutions to such equations will always be of the form

$$x(t) = e^{i\nu t} P(t) \tag{7.11}$$

where $P(t)$ is a periodic function of t, with the same period as the coefficients in Eq. (7.10), and ν is called the characteristic exponent and is, in general, complex. When we have $a_r < p < b_{r+1}$, ν is purely real and, hence, leads to bounded solutions which are aperiodic, in general. And in the opposite case, when $b_r < p < a_{r+1}$, ν is purely imaginary and, hence, leads to unbounded solutions that grow exponentially to infinity.

Equation (7.10) is a second order linear differential equation and thus, has two linearly independent solutions. All other solutions of this equation can be written as a linear superimposition of these two solutions. Writing $P(t)$ as a Fourier series and taking the real part of Eq. (7.11), we can write these two linearly independent solutions as

$$\phi(t) = \sum_{r=-\infty}^{\infty} c_r \cos{(\nu + r\omega)\, t}$$

$$\psi(t) = \sum_{r=-\infty}^{\infty} s_r \sin{(\nu + r\omega)\, t} \tag{7.12}$$

where c_r and s_r are real constants that depend on p, q, ω. Substituting $\phi(t)$ from Eq. (7.12) into Eq. (7.10), and equating coefficients of the same frequency, we get

$$\left[p - (\nu + r\omega)^2\right] c_r = q\left[c_{r+1} + c_{r-1}\right] \tag{7.13}$$

which is a recurrence relation for the coefficients c_r. Carrying out a similar exercise for $\psi(t)$ gives the same recurrence relation as Eq. (7.13) and, thus, $s_r = c_r$.

A one sided infinite sequence of coefficients $(w_0, w_1, w_2, \ldots)$ can be solved by starting from the lowest order term, w_0, and then to keep solving for the coefficients of higher order one by one. However, the coefficients, c_r and s_r, form a two sided infinite sequence and, hence, a different approach needs to be adopted. This is because if we start from the lowest order term, c_0, and proceed in the two directions, the sequence will blow up and not converge. Instead, to approximately solve such recurrence relations, one of the approaches is to artificially truncate the sequence at some $r = \pm R$ and move towards $r = 0$ from the two opposite directions. It is important to note that these two directions yield two different values for c_0. The obtained values of c_r then need to be normalized so as to yield $c_0 = 1$. Obviously, the value of $r = \pm R$ where the series should be truncated depends on the desired accuracy. For values of q which are usually used in Paul trap experiments, $R = 8$ or even $R = 4$, yields good results. And in most cases, the co-efficients in the expression for ϕ and ψ that really matter are c_0, c_1 and c_{-1}. From Eq. (7.13), for $r \gg 0$, $|c_r/c_0| \approx \mathcal{O}(q^{|r|}/4r^2)$.

To be able to solve Eq. (7.13), we also need to know the value of ν. There are two ways of doing this. One is to use a method of continued fractions which is, however, very cumbersome. For low enough values of p, q, it is preferable to use perturbation techniques like the Lindstedt-Poincare method [6], which gives,

$$\nu = \sqrt{p + \frac{2q^2}{\omega^2}} \tag{7.14}$$

As can be seen in Figure 7.6, the particle trajectory is a low frequency sinusoid (both cosine and sine components) with frequency equal to ν, which is irrationally related to ω in general, and two low-amplitude high frequency components, of frequency $w + \nu$ and $w - \nu$, superimposed on that and other higher frequencies as can be seen from the expressions in Eq. (7.12). The low frequency path is the one predicted by the ponderomotive force expression as described in Section 3.4.

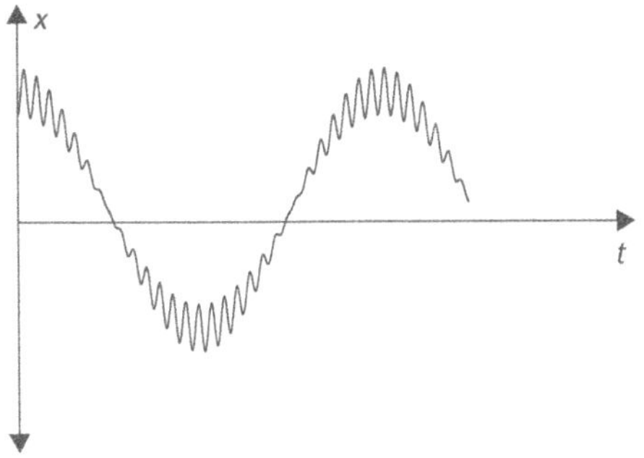

Fig. 7.6: A typical trajectory of a particle along the x-direction in Paul traps.

Though efficient trapping of charged particles had been achieved through the work of W. Paul and others in the 1950s and 60s, trapping of neutral particles was still a major challenge since these particles don't couple to usual electromagnetic fields. In the 1970s and 80s, a very interesting technique was proposed and demonstrated for trapping of neutral particles that exploited its dielectric properties. Neutral particles have no net charge and hence can't directly couple to electromagnetic fields. However, in the presence of EM fields, these particles get polarised, i. e., their electron clouds get distorted leading to a separation of the charge centers. This leads to the emergence of a dipole moment which can experience a net force due to the same EM field which was responsible for the polarisation. And it has now been established that this force can be used to trap neutral particles. This device is known as *optical tweezer* [26, 27] and Steven Chu was awarded the Physics Nobel Prize in 1997 along with Claude Cohen-Tannoudji and William D. Philips for its experimental demonstration.

As shown in Figure 7.7, an optical tweezer essentially consists of a laser beam travelling along the $+z$ direction but whose waist undergoes a focussing effect. The beam width gradually narrows down to reach its minimum (half the laser wavelength) near the origin and then starts increasing again to its original size as the beam moves further along the $+z$ direction. A particle placed on the path of a laser beam experiences the radiation pressure due to which it is forced to move along the direction of propagation of the beam. Hence, a simple laser beam is not sufficient to trap a particle. The stabilising effect is provided by the gradient forces generated due to the focussing effect. Interestingly, the gradient forces act on the dipole so as to push it towards the region of maximum EM intensity. In order to see this, let us now work out the electromagnetic force on an electric dipole created due to the effect of polarisation.

Consider a dipole with charges of magnitude $\pm Q$ placed at the locations $\vec{r}_1$ and $\vec{r}_2$ respectively, where the distance $|\vec{r}_2 - \vec{r}_1|$ will be assumed to be infinitesimal. The total Lorentz force acting on the dipole is given by

$$\vec{F} = Q\left(\vec{E}\left(\vec{r_1}\right) - \vec{E}\left(\vec{r_2}\right) + \frac{d\left(\vec{r}_1 - \vec{r}_2\right)}{dt} \times \vec{B}\right) \tag{7.15}$$

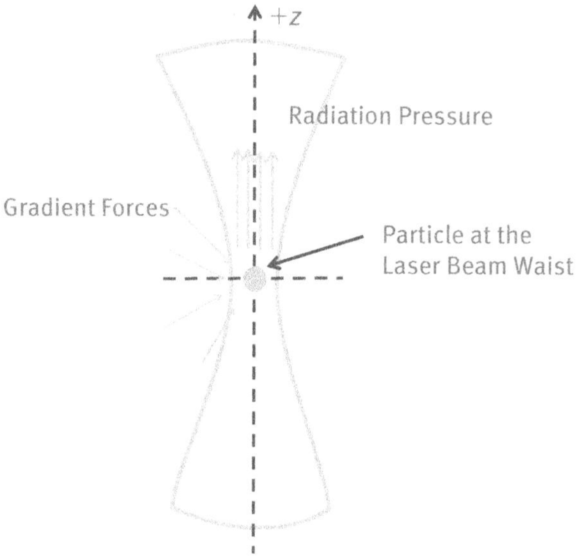

Fig. 7.7: Schematic of an optical tweezer.

Since $\left|\vec{r}_2 - \vec{r}_1\right|$ has been assumed to be small, we can do a Taylor expansion of $\vec{E}\left(\vec{r_2}\right)$ about the point $\vec{r}_1$ to get

$$\vec{E}\left(\vec{r_2}\right) = \vec{E}\left(\vec{r_1}\right) + \left[\left(\vec{r}_2 - \vec{r}_1\right) \cdot \vec{\nabla}\right] \vec{E}\left(\vec{r}_1\right)$$

Substituting this in Eq. (7.15) gives,

$$\vec{F} = \left(\vec{p} \cdot \vec{\nabla}\right) \vec{E}\left(\vec{r}_1\right) + \frac{d\vec{p}}{dt} \times \vec{B} \tag{7.16}$$

where $\vec{p} = Q\left(\vec{r}_1 - \vec{r}_2\right)$ is the dipole moment. Since our dipole moment is created due to the applied electric field, we can write $\vec{p} = \alpha\vec{E}$, where α is the polarisability of the atom and can in general depend on $\vec{E}$, specially for high field intensities. Substituting this in Eq. (7.16), we get

$$\vec{F} = \alpha\left(\left(\vec{E} \cdot \vec{\nabla}\right) \vec{E} + \frac{\partial \vec{E}}{\partial t} \times \vec{B}\right) \tag{7.17}$$

Using vector identities and Faraday's law, we get

$$\begin{aligned} \left(\vec{E} \cdot \vec{\nabla}\right) \vec{E} &= \frac{1}{2}\vec{\nabla}E^2 - \vec{E} \times \left(\vec{\nabla} \times \vec{E}\right) \\ &= \frac{1}{2}\vec{\nabla}E^2 + \vec{E} \times \frac{\partial \vec{B}}{\partial t} \end{aligned}$$

Substituting this in Eq. (7.17), we get

$$\vec{F} = \alpha\left(\frac{1}{2}\vec{\nabla}E^2 + \vec{E}\times\frac{\partial\vec{B}}{\partial t} + \frac{\partial\vec{E}}{\partial t}\times\vec{B}\right)$$
$$= \alpha\left(\frac{1}{2}\vec{\nabla}E^2 + \frac{\partial\left(\vec{E}\times\vec{B}\right)}{\partial t}\right)$$

This force expression is time-varying, but what is of interest in most applications is the time-averaged force experienced by the particle. It can be shown that the 2nd term in this expression averages out to zero over a time-range much larger than the time-period of oscillations of the laser. Hence, the time-averaged force of the laser beam on the dipole is given by

$$\vec{F} = \alpha\frac{1}{2}\vec{\nabla}E^2 \tag{7.18}$$

and is also known as the *gradient force*. A comparison with Eq. (3.25) shows that its form is very similar that of the ponderomotive force (which is responsible for the working of Paul traps) with the only difference of the sign. The ponderomotive force is towards the lower intensity of the electric field and this laser gradient force is towards the higher intensity. Thus, in Figure 7.7, this laser gradient force forces the particle towards the center of the focussing spot (which has the highest intensity in the beam) and balances the radiation pressure that tries to move the particle towards the $+z$ direction.

Though optical tweezers have been very successful in trapping and manipulation of particles for many interesting applications, they are also quite difficult to produce and maintain since they require the operation of very powerful laser beams. Over the last 1990s and early 2000s, it has been proposed and demonstrated that instead of directly using laser beams to trap particles, one could also use evanescent waves and even surface plasmon waves [28, 29]. These *plasmonic tweezers* work on the same principle as optical tweezers but instead of trying to produce a focussed laser beam, they trap particles by coupling of the dipole moment of the dielectric particle with the surface plasmon wave. Like the laser beam in optical tweezers, this surface plasmon (SP) wave also applies two forces on the dielectric particle: radiation pressure along the direction of propagation of the SP wave and a gradient force that moves the particle towards the region of highest plasmon intensity. The main advantage of these plasmonic tweezers is that they can provide very stable trapping of particles with a laser intensity that is two orders of magnitude lower than what is required in optical tweezers. These plasmonic traps also provide very important functionalities for lab-on-the-chip applications at the nanoscale.

8 Mathematical Foundations

In this chapter, we will cover the mathematical concepts required for dealing with the Maxwell's equations and the interaction of electromagnetic fields with charged particles. This chapter has been placed towards the end of this book since we not only discuss the mathematical concepts but also its relation to the electromagnetic concepts discussed in earlier chapters. We will not go into too much details of these concepts and discuss only what is necessary. Readers who might have already learnt these mathematical concepts earlier are also encouraged to go through this chapter since the connections we make with electromagnetics might be new to them. In certain sections of this chapter, we also introduce a few advanced electromagnetic concepts (that are not mentioned earlier in this book) that may be of interest to the reader.

8.1 Scalars and Vectors

Most of the quantities related to electromagnetics or other physical phenomenon can be divided into two categories: those that are just numbers and those that have a sense of direction. *Scalars* are quantities that only have a magnitude and *vectors* are quantities that have both magnitude and direction. The electric potential is a scalar and the electric field is a vector. Since we know that the electric potential and electric field are deeply interconnected, it obviously follows that scalars and vectors are not independent quantities in general. It is possible to derive vector quantities from scalar quantities and *vice-versa*. And going one step further, under certain conditions, a vector can be defined purely by a scalar quantity and *vice-versa*.

8.1.1 Coordinate Systems: Cartesian, Cylindrical and Spherical

Defining a scalar mathematically is very easy since it only has a magnitude. But a vector has a direction and hence its mathematical definition or representation is a bit more tricky. How do we represent a quantity which has a direction? When someone asks us for the directions to a place in the city, we usually point towards certain roads which are fixed. For example, we could ask the person to go straight on the road till the second signal, then take a left and then take a right at the second cut. If the roads of our city were always changing directions, it would have been very difficult to give directions to anyone and the only recourse would perhaps be a GPS based map that keeps updating itself quite frequently. Similarly, to define a vector in space, we need a system of roads that are fixed and do not change with time. How many such roads do we need? A city may need hundreds of roads going in different directions, but thankfully physics can do with only three which are also called the *dimensions* of space.

https://doi.org/10.1515/9783110570038-109

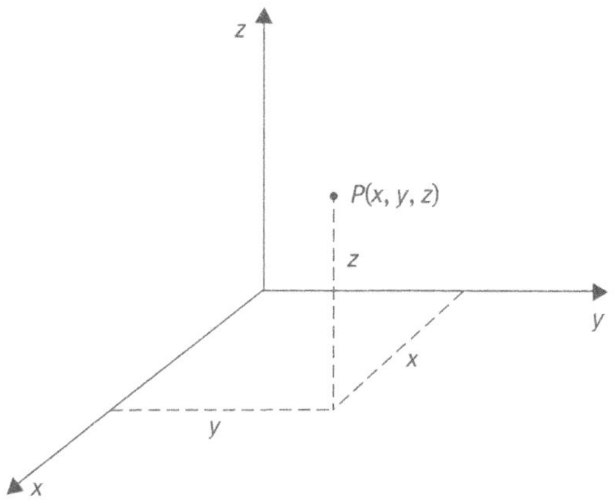

Fig. 8.1: Cartesian coordinate system.

Strictly speaking, modern physics needs 4 dimensions (3 for space and 1 for time), but in this section we will not consider the time dimension.

From our ordinary experience, we also know that it is much easier to find directions if the roads are straight and not curved. So, the simplest map in 3-dimensional (or, 3D) space would consist of 3 straight roads going in different (non-parallel) directions and extending all the way up to infinity both ways (forward and backward). As we will see later, the mathematical analysis of vectors gets immensely simplified if these directions are taken to be perpendicular to each other. Such a coordinate system is said to be *orthogonal*. This simplest coordinate system that is usually used for analysing most of the physical phenomenon is known as the *cartesian* coordinate system and shown in Figure 8.1. This coordinate system consists of three axes labeled as x, y and z. A point in this coordinate system is identified through three numbers (or scalars) corresponding to each of these three axes. Each pair of these three coordinates forms what is known as a *plane* to which the third axis is perpendicular. As shown in Figure 8.1, the z coordinate of the point P is given by the length of the perpendicular drawn from this point on to the $x - y$ plane. If the point P was on the other side of the $x - y$ plane, then its z coordinate would be negative. Similarly, we can find the x and y coordinates of any point in 3D space.

There are two ways to represent this point P in 3D space. One is as a 3-tuple as shown in Figure 8.1 or using the vector notation. In the vector notation, a point in 3D space is written as the sum of three vectors, each of which points along one of the 3 axes,

$$\vec{P} = x\hat{x} + y\hat{y} + z\hat{z} \tag{8.1}$$

where $\hat{x}$ is a unit vector pointing along the $+x$ direction and $\hat{y}$, $\hat{z}$ are defined similarly. The vector $\vec{P}$ is known as the *position vector* of the point P. A unit vector is a vector that has a magnitude (or length) equal to one (unity). In this notation, the coefficient of $\hat{x}$, $\hat{y}$ and $\hat{z}$ cannot be added together and have to be written separately. The length of this vector is denoted by $\left|\vec{P}\right|$ and is given by

$$\left|\vec{P}\right| = \sqrt{x^2 + y^2 + z^2}$$

Though the cartesian coordinate system is very useful in most situations, there are some cases where it can be very cumbersome. An example could be the simple problem of solving for the magnetic field generated due to a cylindrical current carrying wire, as shown in Figure 8.2. From symmetry considerations, we can say that the magnetic field will be the same at all points on a circle whose center is the axis of the wire. Hence, our mathematical analysis of this problem can become much simpler if the distance from the wire axis can be considered to be one of the three coordinates. Such a coordinate system is shown in Figure 8.3 and is known as the *cylindrical* polar coordinates. In this system, one coordinate is z (same as that of cartesian system), one

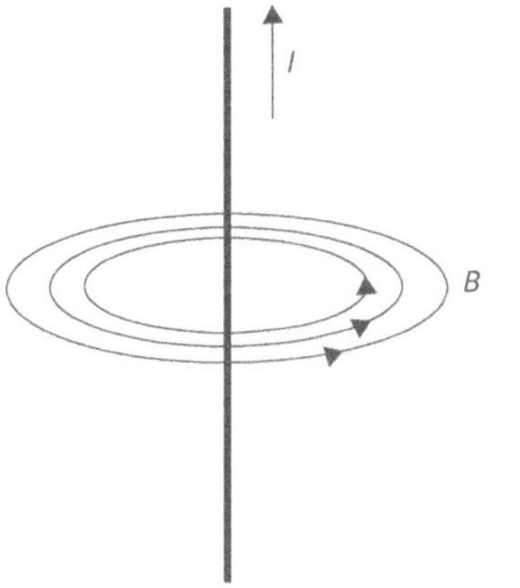

Fig. 8.2: Magnetic field due to a current carrying wire.

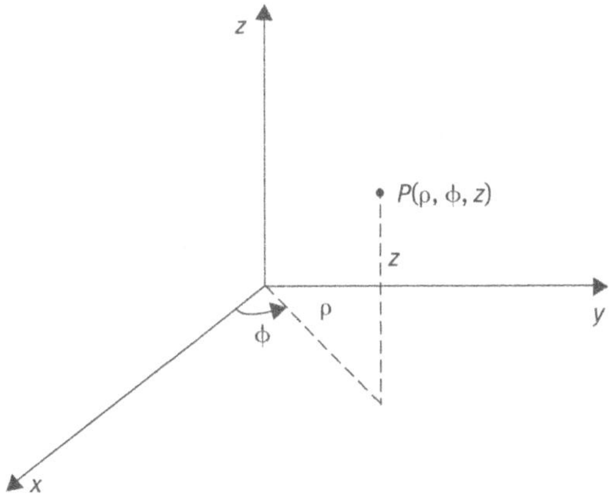

Fig. 8.3: Cylindrical polar coordinate system.

is ρ (distance from z-axis) and one is ϕ (azimuthal angle). Obviously, it is always possible to convert the coordinates of any point from one coordinate system into another one. The following relations can be used to convert the coordinates from the cartesian to the cylindrical system,

$$\rho = \sqrt{x^2 + y^2}$$
$$\phi = \tan^{-1}\frac{y}{x}$$

and these relations can be used to go from the cylindrical to the cartesian system,

$$x = \rho\cos\phi$$
$$y = \rho\sin\phi$$

Here it is important to note that though x, y, z can take any value in the range $[-\infty, \infty]$, the values of ρ, ϕ are restricted. From the above expressions, it follows that $\rho \in [0, \infty]$ and $\phi \in [0, 2\pi]$.

In vector notation in this cylindrical coordinate system, the point P can be written as

$$\vec{P} = \rho\hat{\rho} + z\hat{z} \tag{8.2}$$

where $\hat{\rho} = \cos\phi\hat{x} + \sin\phi\hat{y}$ is the unit vector along the direction of ρ. It is important to note that in this case, the vector notation of a point in space does not have any component along the $\hat{\phi}$ vector since the $\hat{\rho}$ has information about this angle. However, a general vector in this coordinate system will have all the three components,

$$\vec{E} = E_\rho\hat{\rho} + E_\phi\hat{\phi} + E_z\hat{z} \tag{8.3}$$

and its magnitude is given by

$$\left|\vec{E}\right| = \sqrt{E_\rho^2 + E_\phi^2 + E_z^2}$$

Like the cartesian coordinate system, the three unit vectors of the cylindrical system are also orthogonal (perpendicular to each other). However, they are not the same at all points in space. For example, at a point on the x-axis, the $\hat{\rho}$ vector points along $\hat{x}$ and at a point on the y-axis, the $\hat{\rho}$ vector points along $\hat{y}$. Such coordinate systems which change directions at different points in space are called curvilinear coordinate systems. The transformation of a vector $\vec{E}$ from the cartesian to the cylindrical system can be written in the following matrix form,

$$\begin{bmatrix} E_\rho \\ E_\phi \\ E_z \end{bmatrix} = \begin{bmatrix} \cos\phi & \sin\phi & 0 \\ -\sin\phi & \cos\phi & 0 \\ 0 & 0 & 1 \end{bmatrix} \begin{bmatrix} E_x \\ E_y \\ E_z \end{bmatrix} \tag{8.4}$$

and the reverse transformation can be obtained by inverting the 3×3 matrix above,

$$\begin{bmatrix} E_x \\ E_y \\ E_z \end{bmatrix} = \begin{bmatrix} \cos\phi & -\sin\phi & 0 \\ \sin\phi & \cos\phi & 0 \\ 0 & 0 & 1 \end{bmatrix} \begin{bmatrix} E_\rho \\ E_\phi \\ E_z \end{bmatrix} \tag{8.5}$$

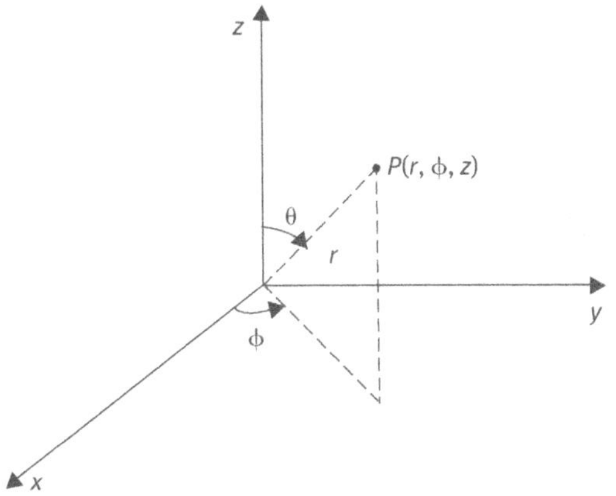

Fig. 8.4: Spherical polar coordinate system.

Another widely used example of a curvilinear coordinate system is the spherical polar coordinate system, as shown in Figure 8.4. This system is very useful when the system under consideration has spherical symmetry. One simple example is the standard problem of finding the electric field due to a spherically symmetric charge density. In this system, the three coordinates of a point are given by r (distance from the origin), ϕ (azimuthal angle, same as that of cylindrical coordinate system) and θ (polar angle measured from the z-axis). The following relations can be used to convert the coordinates from cartesian system to the spherical system,

$$r = \sqrt{x^2 + y^2 + z^2}$$
$$\phi = \tan^{-1}\frac{y}{x}$$
$$\theta = \cos^{-1}\frac{z}{r}$$

and these relations can be used to convert the coordinates from spherical to cartesian coordinates

$$x = r\sin\theta\cos\phi$$
$$y = r\sin\theta\sin\phi$$
$$z = r\cos\theta$$

It follows that the values of these coordinates are restricted to be in these ranges: $r \in [0, \infty]$, $\phi \in [0, 2\pi]$ and $\theta \in [0, \pi]$.

A point in the spherical coordinate system is represented simply by

$$\vec{P} = r\hat{r} \tag{8.6}$$

where $\hat{r}$ is the unit vector from the origin to the point under consideration.

However, a general vector has all the three components,

$$\vec{E} = E_r \hat{r} + E_\phi \hat{\phi} + E_\theta \hat{\theta} \tag{8.7}$$

and its magnitude is given by

$$\left|\vec{E}\right| = \sqrt{E_r^2 + E_\phi^2 + E_\theta^2}$$

The transformation of a vector $\vec{E}$ from the cartesian to the spherical system can be written in the following matrix form,

$$\begin{bmatrix} E_r \\ E_\phi \\ E_\theta \end{bmatrix} = \begin{bmatrix} \sin\theta\cos\phi & \sin\theta\sin\phi & \cos\theta \\ \cos\theta\cos\phi & \cos\theta\sin\phi & -\sin\theta \\ -\sin\phi & \cos\phi & 0 \end{bmatrix} \begin{bmatrix} E_x \\ E_y \\ E_z \end{bmatrix} \tag{8.8}$$

and the reverse transformation can be obtained by inverting the 3×3 matrix above,

$$\begin{bmatrix} E_x \\ E_y \\ E_z \end{bmatrix} = \begin{bmatrix} \sin\theta\cos\phi & \cos\theta\cos\phi & -\sin\phi \\ \sin\theta\sin\phi & \cos\theta\sin\phi & \cos\phi \\ \cos\theta & -\sin\theta & 0 \end{bmatrix} \begin{bmatrix} E_\rho \\ E_\phi \\ E_z \end{bmatrix} \tag{8.9}$$

A very useful parameter in electromagnetics is the distance between two points, $P_1(x_1, y_1, z_1)$ and $P_2(x_2, y_2, z_2)$ and is given by

$$\begin{aligned} d^2 &= \left|\vec{P}_2 - \vec{P}_1\right|^2 \qquad (8.10) \\ &= (x_2 - x_1)^2 + (y_2 - y_1)^2 + (z_2 - z_1)^2 \qquad \text{Cartesian} \\ &= \rho_1^2 + \rho_2^2 - 2\rho_1\rho_2\cos(\phi_2 - \phi_1) + (z_2 - z_1)^2 \qquad \text{Cylindrical} \\ &= r_1^2 + r_2^2 - 2r_1r_2\cos\theta_1\cos\theta_2 \\ &\quad - 2r_1r_2\sin\theta_1\sin\theta_2\cos(\phi_2 - \phi_1) \qquad \text{Spherical} \end{aligned}$$

8.1.2 Gradient of a Scalar

The scalars and vectors defined in the previous section are in general functions of the coordinate variables. And a quantity that plays a very important role in electromagnetics and in all of physics, in general, is the rate of change of these scalar and vector functions with respect to the coordinates. In 1D space, this rate of change is easy to calculate and is given by the derivative of the function with respect to the coordinate variable. However, in 3D space, one has to be a little more careful since to define the rate of change of a function, we now also need to specify the direction along which we are interested in finding the rate. In cartesian coordinate system, the rate of change of a scalar function, V, along the $+x$ direction is given by $\partial V/\partial x$. Similarly, the rate of change along the $+y$ direction is given by $\partial V/\partial y$ and that along the $+z$ direction is given by $\partial V/\partial z$. The partial derivative, $\partial/\partial x$, denotes the fact that other variables are kept constant when the derivative with respect to x is being taken. This is very important in electromagnetic theory since all functions in this case are generally functions of more than one variable. Hence, the rate of change also has three components and can be written in the vector form,

$$\begin{aligned}\vec{\nabla} V &= \frac{\partial V}{\partial x}\hat{x} + \frac{\partial V}{\partial y}\hat{y} + \frac{\partial V}{\partial z}\hat{z} && \text{Cartesian}\\ &= \frac{\partial V}{\partial \rho}\hat{\rho} + \frac{1}{\rho}\frac{\partial V}{\partial \phi}\hat{\phi} + \frac{\partial V}{\partial z}\hat{z} && \text{Cylindrical}\\ &= \frac{\partial V}{\partial r}\hat{r} + \frac{1}{r}\frac{\partial V}{\partial \theta}\hat{\theta} + \frac{1}{r\sin\theta}\frac{\partial V}{\partial \phi}\hat{\phi} && \text{Spherical}\end{aligned} \tag{8.11}$$

where

$$\vec{\nabla} = \frac{\partial}{\partial x}\hat{x} + \frac{\partial}{\partial y}\hat{y} + \frac{\partial}{\partial z}\hat{z}$$

is known as the *gradient operator*. In electrostatics, the electric field vector can be written as the negative gradient of the scalar potential, $\vec{E} = -\vec{\nabla} V$. Hence, the electric field at a certain point in space is a measure of the spatial rate at which the electric potential is changing at that point. The term 'rate' is usually used with respect to time and so care must be taken not to confuse the two concepts. The rate of change of a function with respect to time is denoted by $\partial/\partial t$ or d/dt depending on whether we are interested in taking the partial or total derivative.

8.1.3 Divergence and Curl of a Vector

The gradient operator defines the rate of change of a scalar in 3D space. Similarly, we can also define an operator to measure the rate of change of a vector in 3D space. But as it turns out, this can be done in two ways. Consider a vector field, $\vec{E}(x, y, z)$, and a certain point, $P(x, y, z)$ within the region in space where our vector function is defined. When we want to measure the rate of change of a vector field, it is important to note that the change may be due to the magnitude of the function or its direction. As we move in the neighbourhood of the point, P, broadly speaking, there are two ways in which our vector can change directions. As shown in Figure 8.5, the vector field can either be point away from (or towards) the point under consideration in all directions or, it can be going around the point under consideration in spirals. All other possibilities are a superposition of these two broad categories. The *divergence* of a vector field measures the rate at which the field is pointing away or towards a given point in space. And the *curl* of a vector field measures the rate at which it spirals around the point. The divergence of a vector field, $\vec{E}$, is given by

$$\vec{\nabla} \cdot \vec{E} = \frac{\partial E_x}{\partial x} + \frac{\partial E_y}{\partial y} + \frac{\partial E_z}{\partial z} \qquad \text{Cartesian}$$

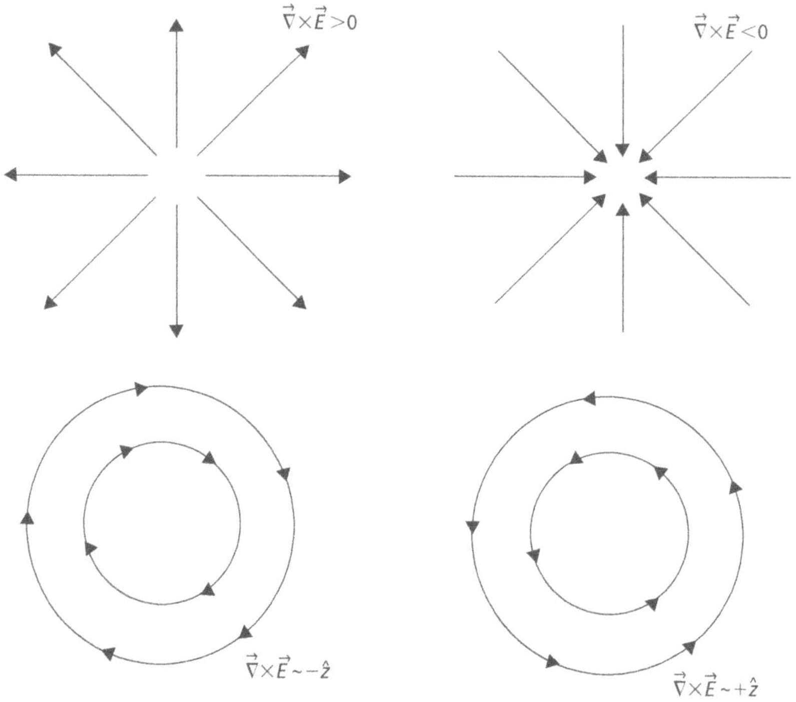

Fig. 8.5: Change of vector field around a certain point in the $x-y$ plane.

$$\begin{aligned}\vec{\nabla}\cdot\vec{E} &= \frac{1}{\rho}\frac{\partial(\rho E_\rho)}{\partial\rho}+\frac{1}{\rho}\frac{\partial E_\phi}{\partial\phi}+\frac{\partial E_z}{\partial z} \qquad \text{Cylindrical}\\ &= \frac{1}{r^2}\frac{\partial}{\partial r}\left(r^2E_r\right)+\frac{1}{r\sin\theta}\frac{\partial}{\partial\theta}(\sin\theta E_\theta)+\frac{1}{r\sin\theta}\frac{\partial E_\phi}{\partial\phi} \qquad \text{Spherical}\end{aligned} \tag{8.12}$$

and the curl is given by

$$\begin{aligned}\vec{\nabla}\times\vec{E} &= \left(\frac{\partial E_z}{\partial y}-\frac{\partial E_y}{\partial z}\right)\hat{x}+\left(\frac{\partial E_x}{\partial z}-\frac{\partial E_z}{\partial x}\right)\hat{y}\\ &\quad+\left(\frac{\partial E_y}{\partial x}-\frac{\partial E_x}{\partial y}\right)\hat{z} \qquad \text{Cartesian}\\ &= \left[\frac{1}{\rho}\frac{\partial E_z}{\partial\phi}-\frac{\partial E_\phi}{\partial z}\right]\hat{\rho}+\left[\frac{\partial E_\rho}{\partial z}-\frac{\partial E_z}{\partial\rho}\right]\hat{\phi}\\ &\quad+\frac{1}{\rho}\left[\frac{\partial(\rho E_\phi)}{\partial\rho}-\frac{\partial E_\rho}{\partial\phi}\right]\hat{z} \qquad \text{Cylindrical}\\ &= \frac{1}{r\sin\theta}\left[\frac{\partial}{\partial\theta}(\sin\theta E_\phi)-\frac{\partial E_\theta}{\partial\phi}\right]\hat{r}+\frac{1}{r}\left[\frac{1}{\sin\theta}\frac{\partial E_r}{\partial\phi}-\frac{\partial}{\partial r}(rE_\phi)\right]\hat{\theta}\\ &\quad+\frac{1}{r}\left[\frac{\partial}{\partial r}(rE_\theta)-\frac{\partial E_r}{\partial\theta}\right]\hat{\phi} \qquad \text{Spherical}\end{aligned} \tag{8.13}$$

The divergence and the gradient operator can be combined to give what is known as the *Laplace operator*,

$$\nabla^2 V = \vec{\nabla} \cdot \left(\vec{\nabla} V\right) \tag{8.14}$$

which plays a very important role in electrostatics and many other physical phenomenon. For a charge free space, the Laplacian of the scalar potential is zero. In the various coordinate systems, this operator can be written as

$$\begin{aligned}
\nabla^2 V &= \frac{\partial^2 V}{\partial x^2} + \frac{\partial^2 V}{\partial y^2} + \frac{\partial^2 V}{\partial z^2} && \text{Cartesian} \\
&= \frac{1}{\rho}\frac{\partial}{\partial \rho}\left(\rho \frac{\partial V}{\partial \rho}\right) + \frac{1}{\rho^2}\frac{\partial^2 V}{\partial \phi^2} + \frac{\partial^2 V}{\partial z^2} && \text{Cylindrical} \\
&= \frac{1}{r^2}\frac{\partial}{\partial r}\left(r^2 \frac{\partial V}{\partial r}\right) + \frac{1}{r^2 \sin\theta}\frac{\partial}{\partial \theta}\left(\sin\theta \frac{\partial V}{\partial \theta}\right) + \frac{1}{r^2 \sin^2\theta}\frac{\partial^2 V}{\partial \phi^2} && \text{Spherical}
\end{aligned}$$

The Laplace equation, $\nabla^2 V = 0$, can be solved analytically only for simple geometries by using the separation of variables method. For practical purposes, it has to be solved numerically using techniques described in Chapter 2.

8.1.4 Scalar and Vector Integration

In high school mathematics, integration is usually defined as the area under a curve. Though this is easy to understand for the case of 1D space, the situation becomes more complex in 2D and 3D space where the direction of integration also has to be defined along with the limits. For example, consider a function in 2D space, $f(x, y)$, which needs to be integrated from $P_1(x_1, y_1)$ to $P_2(x_2, y_2)$, as shown in Figure 8.6. As can be seen in this figure, there are multiple ways of joining these two points and the result of the integration will, in general, depend on the *path* chosen. This holds true even if the function to be integrated is a vector. In addition, the integration of a vector function along a given path between two points can be carried out in multiple ways.

Consider a vector function in 3D space, $\vec{E}(x, y, z)$, which is to be integrated from points $P_1(x_1, y_1, z_1)$ to $P_2(x_2, y_2, z_2)$ along a given path. We can either integrate each component of the vector $\vec{E}$ along this path and the final result will be another vector formed by the results of each of the three integrations. Or, we can take a product of the vector $\vec{E}$ with an infinitesimal vector element along the path, $\vec{dl}$. This line element $\vec{dl}$ is tangential at every point to the path drawn between P_1 and P_2. There are two ways of multiplying two vectors. One is the *dot product* and another is known as the *cross product*. For any two given vectors, $\vec{E}$ and $\vec{B}$, the dot product is a scalar and simply given by

$$\vec{E} \cdot \vec{B} = E_x B_x + E_y B_y + E_z B_z \tag{8.15}$$

For the same two vectors, the cross product is a vector and given by

$$\vec{E} \times \vec{B} = (E_y B_z - E_z B_y)\,\hat{x} + (E_z B_x - E_x B_z)\,\hat{y} + (E_x B_y - E_y B_x)\,\hat{z} \tag{8.16}$$

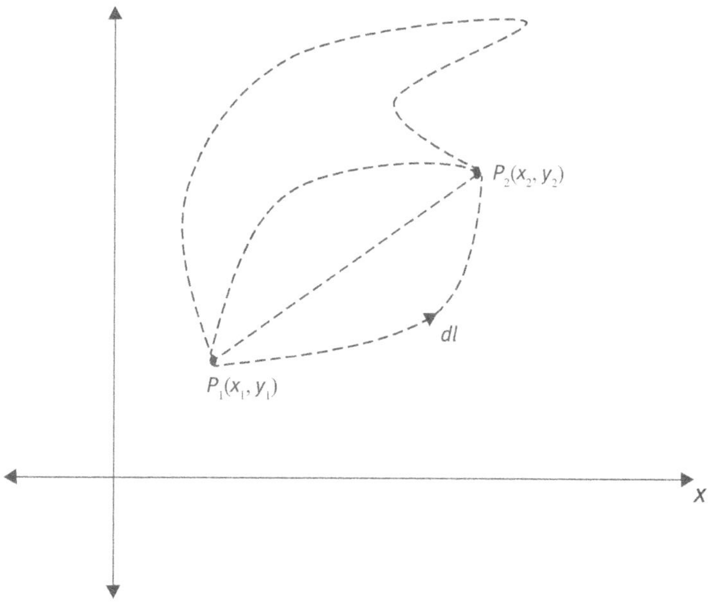

Fig. 8.6: Integration paths on a 2D plane.

It obviously follows from these equations that

$$\vec{E} \cdot \vec{B} = \vec{B} \cdot \vec{E}$$
$$\vec{E} \times \vec{B} = -\vec{B} \times \vec{E}$$

It can also be checked that the direction of $\vec{E} \times \vec{B}$ is perpendicular to both $\vec{E}$ and $\vec{B}$, and is given by the right hand thumb rule. Thus, one can write $\hat{x} = \hat{y} \times \hat{z}$, $\hat{y} = \hat{z} \times \hat{x}$ and $\hat{z} = \hat{x} \times \hat{y}$ for the unit vectors in cartesian coordinates. Similar expressions can be written for the unit vectors in cylindrical and spherical coordinate systems.

So if we take a cross product of $\vec{E}$ and $\vec{dl}$, we will get another vector whose each component can be integrated to yield the final result of the integration. However, when we take a dot product of $\vec{E}$ and $\vec{dl}$, we get a scalar whose integration yields another scalar. This integration of $\vec{E} \cdot \vec{dl}$ is known as a *line integral*. In the three coordinate systems mentioned above, the line element is given by

$$\begin{aligned} \vec{dl} &= dx\hat{x} + dy\hat{y} + dz\hat{z} \qquad \text{Cartesian} \\ &= d\rho\hat{\rho} + \rho d\phi\hat{\phi} + dz\hat{z} \qquad \text{Cylindrical} \\ &= dr\hat{r} + rd\theta\hat{\theta} + r\sin\theta d\phi\hat{\phi} \qquad \text{Spherical} \end{aligned} \tag{8.17}$$

There is a special case when the line integral of a vector function does not depend on the path of the integration and is completely determined by the end points. This happens when the vector function can be written as the gradient of a scalar function, which we know holds true in electrostatics. If the electric and magnetic fields in a certain region of space are time-independent, then the electric field can be written as the negative gradient of the electric potential (V),

$$\vec{E} = -\vec{\nabla} V$$

In this case, the line integral becomes

$$\begin{aligned}\int_{P_1}^{P_2} \vec{E} \cdot \vec{dl} &= -\int_{P_1}^{P_2} \left(\vec{\nabla} V\right) \cdot \vec{dl} \\ &= V(P_1) - V(P_2) \end{aligned} \tag{8.18}$$

which follows from *gradient theorem* or the *fundamental theorem of calculus* for line integral. Thus, this theorem says that if there is a field in space which can be defined as the gradient of a scalar, V, then its line integral is simply given by the difference of the scalar at the end points of the curve. Such fields are known as conservative fields. If a region of space contains time-varying magnetic fields, then the electric field in that region is no longer conservative and this is the origin of Faraday's law of induction.

Like we have defined the line integral, we can also define the surface and volume integral. Like the path of a line integral is bounded by two points, the surface integral is over a surface bounded by a closed curve and the volume integral is over a volume enclosed by a closed surface. The surface element is also a vector denoted by $\vec{dS}$ and is tangential at every point on the surface over which the integration is being done. The volume element is, however, a scalar and denoted by $d\tau$,

$$\begin{aligned} d\tau &= dxdydz \qquad \text{Cartesian} \\ &= \rho d\rho d\phi dz \qquad \text{Cylindrical} \\ &= r^2 \sin\theta dr d\theta d\phi \qquad \text{Spherical} \end{aligned}$$

Like the gradient theorem connects a line integral with simple scalar integration, there are two other very important theorems which connect a surface integral to a volume integral and also to a line integral.

The *divergence theorem*, also known as Gauss' theorem or Ostrogradsky's theorem gives

$$\int \left(\vec{\nabla} \cdot \vec{E}\right) d\tau = \oint \vec{E} \cdot \vec{dS} \tag{8.19}$$

This theorem says that the surface integral of a certain field in space, $\vec{E}$, is equal to the volume integral of its divergence. This is very important in electrostatics since we know from Gauss law that $\vec{\nabla} \cdot \vec{E} = \rho_q/\epsilon_0$ (where ρ_q is the charge density and not the coordinate of cylindrical system) and hence the LHS of the above equation becomes Q/ϵ_0 where $Q = \int \rho_q d\tau$ is the total charge enclosed in the volume.

The *curl theorem*, also known as the Stokes' theorem or the Stokes-Cartan theorem gives

$$\int \left(\vec{\nabla} \times \vec{B}\right) \cdot \vec{dS} = \oint \vec{B} \cdot \vec{dl} \tag{8.20}$$

This theorem says that the surface integral of the curl of a certain field in space, $\vec{B}$, is equal to the line integral around a closed loop bounding the surface. This is very useful in magnetostatics since we know from Ampere's law that in the absence of time-varying electric fields, the curl of $\vec{B}$ is proportional to the current density, $\vec{J}$, i. e., we have $\vec{\nabla} \times \vec{B} = \mu_0 \vec{J}$. Hence, the LHS of the above equation reduces to the total charge flowing through the surface under consideration. What this implies is that in order to find the magnetic field along a circular loop shown in Figure 8.2, we do not need information about the detailed current density distribution along the wire. As long as the current density has cylindrical symmetry (depends only on $\vec{r}$) and total current is known, the magnetic field is independent of the detailed distribution.

8.1.5 Vector Identities

For solving the Maxwell's equations, we have to repeatedly do several vector manipulations. Here, it is very useful to know certain vector identities that help in transforming the equations into forms that are easily solvable. Below is a list of such identities and the reader is encouraged to prove each of them:

$$\vec{\nabla}(fg) = f\vec{\nabla}g + g\vec{\nabla}f$$

$$\vec{\nabla}\left(\vec{E}\cdot\vec{B}\right) = \vec{E}\times\left(\vec{\nabla}\times\vec{B}\right) + \vec{B}\times\left(\vec{\nabla}\times\vec{E}\right) + \left(\vec{E}\cdot\vec{\nabla}\right)\vec{B} + \left(\vec{B}\cdot\vec{\nabla}\right)\vec{E}$$

$$\vec{\nabla}\cdot\left(g\vec{E}\right) = g\left(\vec{\nabla}\cdot\vec{E}\right) + \vec{E}\cdot\left(\vec{\nabla}g\right)$$

$$\vec{\nabla}\cdot\left(\vec{E}\times\vec{B}\right) = \vec{B}\cdot\left(\vec{\nabla}\times\vec{E}\right) - \vec{E}\cdot\left(\vec{\nabla}\times\vec{B}\right)$$

$$\vec{\nabla}\times\left(g\vec{E}\right) = g\left(\vec{\nabla}\times\vec{E}\right) - \vec{E}\times\left(\vec{\nabla}g\right)$$

$$\vec{\nabla}\times\left(\vec{\nabla}\times\vec{E}\right) = \vec{\nabla}\left(\vec{\nabla}\cdot\vec{E}\right) - \nabla^2\vec{E}$$

$$\vec{\nabla}\times\left(\vec{E}\times\vec{B}\right) = \left(\vec{B}\cdot\vec{\nabla}\right)\vec{E} - \left(\vec{E}\cdot\vec{\nabla}\right)\vec{B} + \vec{E}\left(\vec{\nabla}\cdot\vec{B}\right) - \vec{B}\left(\vec{\nabla}\cdot\vec{E}\right)$$

$$\vec{A}\cdot\left(\vec{B}\times\vec{C}\right) = \vec{B}\cdot\left(\vec{C}\times\vec{A}\right) = \vec{C}\cdot\left(\vec{A}\times\vec{B}\right)$$

$$\vec{A}\times\left(\vec{B}\times\vec{C}\right) = \vec{B}\left(\vec{A}\cdot\vec{C}\right) - \vec{C}\left(\vec{A}\cdot\vec{B}\right)$$

$$\vec{\nabla}\cdot\left(\vec{\nabla}\times\vec{A}\right) = 0 \tag{8.21}$$

$$\vec{\nabla}\times\left(\vec{\nabla}V\right) = 0 \tag{8.22}$$

8.1.6 Scalar and Vector Potential

The two identities given by Eqs. (8.21) and (8.22) are of particular importance in electromagnetics. What the first one says is that if the divergence of a certain vector is zero, then it can be written as the curl of another vector. And the second one says that if the curl of a vector is zero, then it can be written as the gradient of a scalar. These two identities are used to define the vector and scalar potentials in electromagnetics which in turn, reduce the Maxwell's equations to two in number from the conventional four. These potentials are also very useful in solving for the electromagnetic fields generated due to various charge and current configurations.

According to Gauss' law for magnetic fields, we have

$$\vec{\nabla} \cdot \vec{B} = 0$$

in the absence of magnetic monopoles. Thus, as per Eq. (8.21), we can write $\vec{B}$ as the curl of a vector, $\vec{A}$, known as the vector potential. Thus, if we write

$$\vec{B} = \vec{\nabla} \times \vec{A}, \tag{8.23}$$

we no longer need a separate Gauss' law for magnetic fields since it is now automatically satisfied.

Substituting Eq. (8.23) in Faraday's law, we get

$$\begin{aligned} \vec{\nabla} \times \vec{E} &= -\frac{\partial \vec{B}}{\partial t} \\ &= -\frac{\partial \left(\vec{\nabla} \times \vec{A}\right)}{\partial t} \\ \Rightarrow \vec{\nabla} \times \left(\vec{E} + \frac{\partial \vec{A}}{\partial t}\right) &= 0 \end{aligned}$$

Now using Eq. (8.22), we can say that $\vec{E} + \partial\vec{A}/\partial t$ can be written as the negative gradient of a scalar potential, V. Thus, the electric field is given by

$$\vec{E} = -\vec{\nabla} V - \frac{\partial \vec{A}}{\partial t} \tag{8.24}$$

So now if we write the electric field in this format, we do not need to separately write the Faraday's law since it is automatically taken care of. We are left with only two of the original four Maxwell's equations and it is these two equations that govern the dynamics of V and $\vec{A}$.

Substituting Eqs. (8.23) and (8.24) in the Ampere-Maxwell equation, we get

$$\vec{\nabla} \times \vec{B} = \mu_0 \vec{J} + \mu_0 \epsilon_0 \frac{\partial \vec{E}}{\partial t}$$

$$\Rightarrow \vec{\nabla} \times \left(\vec{\nabla} \times \vec{A}\right) = \mu_0 \vec{J} + \mu_0 \epsilon_0 \frac{\partial}{\partial t}\left(-\vec{\nabla} V - \frac{\partial \vec{A}}{\partial t}\right)$$

$$\Rightarrow \vec{\nabla}\left(\vec{\nabla} \cdot \vec{A}\right) - \nabla^2 \vec{A} = \mu_0 \vec{J} - \mu_0 \epsilon_0 \vec{\nabla} \frac{\partial V}{\partial t} - \mu_0 \epsilon_0 \frac{\partial^2 \vec{A}}{\partial t^2}$$

$$\Rightarrow \nabla^2 \vec{A} - \mu_0 \epsilon_0 \frac{\partial^2 \vec{A}}{\partial t^2} + \vec{\nabla}\left[\vec{\nabla} \cdot \vec{A} + \mu_0 \epsilon_0 \frac{\partial V}{\partial t}\right] = -\mu_0 \vec{J}$$

Since the electric and magnetic fields are defined as derivatives of the scalar and vector potentials, there is some inherent freedom in their choice. We can make a convenient choice for $\vec{A}$ and V which leaves the electric and magnetic fields unchanged, but significantly simplifies the equations governing their dynamics. One such choice known as the Lorentz gauge is

$$\vec{\nabla} \cdot \vec{A} + \mu_0 \epsilon_0 \frac{\partial V}{\partial t} = 0 \tag{8.25}$$

Using this condition, the equation for the vector potential becomes,

$$\nabla^2 \vec{A} - \frac{1}{c^2} \frac{\partial^2 \vec{A}}{\partial t^2} = -\mu_0 \vec{J} \tag{8.26}$$

Now substituting Eq. (8.24) in the Gauss' law for electric field and using the Lorentz gauge from Eq. (8.25), we get

$$\vec{\nabla} \cdot \vec{E} = \frac{\rho_q}{\epsilon_0}$$

$$\Rightarrow \vec{\nabla} \cdot \left(-\vec{\nabla} V - \frac{\partial \vec{A}}{\partial t}\right) = \frac{\rho_q}{\epsilon_0}$$

$$\Rightarrow \nabla^2 V - \frac{1}{c^2} \frac{\partial^2 V}{\partial t^2} = -\frac{\rho_q}{\epsilon_0} \tag{8.27}$$

Equations (8.26) and (8.27) are essentially wave equations with a driving force term provided by the current density and charge density respectively. An important consequence of these wave equations is that they impose a certain finite rate of transfer of information. If a temporal change in the current/charge density happens at a certain point in space, it takes a certain non-zero time for this disturbance to reach another point in space. This speed of transfer of information is also fixed and is given by the speed of light in free space, $c = 3 \times 10^8 m/s$. It is this fundamental idea that lead Albert Einstein to discover his special theory of relativity. These two equations can be solved by writing them in the integral form,

$$V(\vec{r}, t) = \frac{1}{4\pi\epsilon_0} \int \frac{\rho_q\left(\vec{r}', t_r\right)}{\left|\vec{r} - \vec{r}'\right|} d\tau'$$

$$\vec{A}(\vec{r},t) = \frac{\mu_0}{4\pi}\int \frac{\vec{J}\left(\vec{r}',t_r\right)}{\left|\vec{r}-\vec{r}'\right|}d\tau' \tag{8.28}$$

where $t_r = t - \left|\vec{r}-\vec{r}'\right|/c$ is known as the retarded time.

8.2 Lorentz Transformations and Special Relativity

When physicists encounter an equation that describe some fundamental property of this universe, one question they naturally ask is regarding the coordinate transformations that leave these equations invariant. Consider two reference frames (coordinate systems), $S(x,y,z,t)$ and $S'\left(x',y',z',t'\right)$, in 3D space such that S' is moving with respect to S along with a constant velocity, $\vec{u}$. It obviously follows that S is moving with respect to S' with the velocity, $-\vec{u}$. As per the usual rules of *Galilean transformation* commonly used in Newtonian mechanics, we would write,

$$\begin{aligned} \vec{r}' &= \vec{r} - \vec{u}t \\ t' &= t \end{aligned} \tag{8.29}$$

Now let us apply these transformations to the Maxwell's equations in free space and see what we get. Let the electric and magnetic field in S coordinate system be $\vec{E}$, $\vec{B}$ and in the S' coordinate system be $\vec{E}'$, $\vec{B}'$. We first need to find out the relation between the fields in these two coordinates. The force experienced by a charged particle present at a point in space and moving with a velocity $\vec{v}$ in S (or $\vec{v}'$ in S') is given by the Lorentz force expression,

$$\begin{aligned} \vec{F} &= Q\left[\vec{E} + \vec{v}\times\vec{B}\right] \\ \vec{F}' &= Q\left[\vec{E}' + \vec{v}'\times\vec{B}'\right] \\ &= Q\left[\vec{E}' + (\vec{v}-\vec{u})\times\vec{B}'\right] \end{aligned}$$

where we have used $\vec{v}' = \vec{v} - \vec{u}$ as given by the Galilean transformation. Since this force expression does not depend on acceleration and we can see that $\ddot{\vec{r}}' = \ddot{\vec{r}}$ (where $\ddot{}$ represents double derivative with respect to the respective time coordinate), the two force terms must be the same. Hence, it follows that

$$\begin{aligned} \vec{E} + \vec{v}\times\vec{B} &= \vec{E}' + (\vec{v}-\vec{u})\times\vec{B}' \\ \Rightarrow \vec{E} + \vec{v}\times\left(\vec{B}-\vec{B}'\right) &= \vec{E}' - \vec{u}\times\vec{B}' \end{aligned}$$

Since this equation must hold true for any choice of $\vec{v}$ and $\vec{u}$, the only possibility is that

$$\begin{aligned} \vec{B}' &= \vec{B} \\ \vec{E}' &= \vec{E} + \vec{u}\times\vec{B} \end{aligned} \tag{8.30}$$

In order to write the Maxwell's equations in the S' coordinate system, we now need to write the equivalent expressions for the spatial and temporal derivatives in this coordinate system. Using the rules of partial derivatives, it follows that

$$\begin{aligned}\frac{\partial}{\partial x'} &= \frac{\partial t}{\partial x'}\frac{\partial}{\partial t} + \frac{\partial x}{\partial x'}\frac{\partial}{\partial x} + \frac{\partial y}{\partial x'}\frac{\partial}{\partial y} + \frac{\partial z}{\partial x'}\frac{\partial}{\partial z}\\ &= \frac{\partial}{\partial x}\\ \Rightarrow \vec{\nabla}' &= \vec{\nabla}\end{aligned}$$

$$\begin{aligned}\frac{\partial}{\partial t'} &= \frac{\partial t}{\partial t'}\frac{\partial}{\partial t} + \frac{\partial x}{\partial t'}\frac{\partial}{\partial x} + \frac{\partial y}{\partial t'}\frac{\partial}{\partial y} + \frac{\partial z}{\partial t'}\frac{\partial}{\partial z}\\ &= \frac{\partial t}{\partial t'}\frac{\partial}{\partial t} + u_x\frac{\partial}{\partial x} + u_y\frac{\partial}{\partial y} + u_z\frac{\partial}{\partial z}\end{aligned}$$

$$\begin{aligned}\frac{\partial}{\partial t'} &= \frac{\partial}{\partial t} + \left(\vec{u}\cdot\vec{\nabla}\right)\\ \Rightarrow \frac{\partial}{\partial t} &= \frac{\partial}{\partial t'} - \left(\vec{u}\cdot\vec{\nabla}'\right)\end{aligned} \tag{8.31}$$

Applying Eqs. (8.30) and (8.31) to Gauss' law for electric field in free space, we get

$$\begin{aligned}\vec{\nabla}\cdot\vec{E} &= 0\\ \Rightarrow \vec{\nabla}'\cdot\left(\vec{E}' - \vec{u}\times\vec{B}'\right) &= 0\\ \Rightarrow \vec{\nabla}'\cdot\vec{E}' - \vec{\nabla}'\cdot\left(\vec{u}\times\vec{B}'\right) &= 0\end{aligned}$$

Hence, in order for the Gauss' law to have the same form in the S' coordinate system, we must have $\vec{\nabla}'\cdot\left(\vec{u}\times\vec{B}'\right) = 0$, which is not true in general. Hence, some of the Maxwell's equations are not invariant under the Galilean transformation while some of them are. It is easy to see that $\vec{\nabla}\cdot\vec{B} = 0$ is invariant under Galilean transformation since $\vec{\nabla}' = \vec{\nabla}$ and $\vec{B}' = \vec{B}$. We also need to check whether the Galilean transformation leaves the wave equation invariant even if it changes the Maxwell's equations since in most cases it is the wave equation that we are finally interested in. In the S coordinate system, the wave equation in free space is given by

$$\left(\nabla^2 - \frac{1}{c^2}\frac{\partial}{\partial t^2}\right)\Psi = 0 \tag{8.32}$$

Now, from Eq. (8.31), we can see that $\nabla'^2 = \vec{\nabla}'\cdot\vec{\nabla}' = \vec{\nabla}\cdot\vec{\nabla} = \nabla^2$ and hence, the Laplacian operator is invariant under Galilean transformation. The double time derivative is a little more complicated and transforms in the following way,

$$\begin{aligned}\frac{\partial^2}{\partial t^2} &= \frac{\partial}{\partial t}\frac{\partial}{\partial t}\\ &= \left[\frac{\partial}{\partial t'} - \left(\vec{u}\cdot\vec{\nabla}'\right)\right]\left[\frac{\partial}{\partial t'} - \left(\vec{u}\cdot\vec{\nabla}'\right)\right]\\ &\neq \frac{\partial^2}{\partial t'^2}\end{aligned}$$

Thus, the wave equation is clearly not invariant under Galilean transformation. However, as it turns out, the wave equation and all the four Maxwell's equations are invariant under what is known as the *Lorentz transformation*. For the case of simplicity, we will consider the case of 1D wave motion where the wave is propagating along the $+x$ direction and is independent of the y, z coordinates. Also, the reference frame S' is now moving with respect to S along the $+x$ direction with a velocity u_x. Hence, the y, z coordinates of the two coordinate systems do not undergo any change, i. e., $y' = y$ and $z' = z$. The x, t coordinate transformations are now governed by rules that are slightly more complicated than what we had in the case of Galilean transformation,

$$\begin{aligned} x' &= \gamma(x - u_x t) \\ t' &= \gamma\left(t - \frac{u_x x}{c^2}\right) \\ \gamma &= \frac{1}{\sqrt{1 - u_x^2/c^2}} \end{aligned} \tag{8.33}$$

where c is the speed of light in free space. The important point to note is that under the Lorentz transformation, the time coordinate also undergoes a change, which is quite contrary to our intuitive experience of the world where time is assumed to be a universal phenomenon ticking away at a constant rate for everybody. Now let us apply these transformations to the wave equation and see what happens. In order to do that, we first need to find the transformation rules for the derivatives,

$$\begin{aligned} \frac{\partial}{\partial x} &= \frac{\partial x'}{\partial x}\frac{\partial}{\partial x'} + \frac{\partial t'}{\partial x}\frac{\partial}{\partial t'} \\ &= \gamma\frac{\partial}{\partial x'} - \frac{\gamma u_x}{c^2}\frac{\partial}{\partial t'} \\ \Rightarrow \frac{\partial^2}{\partial x^2} &= \frac{\partial}{\partial x}\frac{\partial}{\partial x} \\ &= \left[\gamma\frac{\partial}{\partial x'} - \frac{\gamma u_x}{c^2}\frac{\partial}{\partial t'}\right]\left[\gamma\frac{\partial}{\partial x'} - \frac{\gamma u_x}{c^2}\frac{\partial}{\partial t'}\right] \\ &= \gamma^2\frac{\partial^2}{\partial x'^2} + \frac{\gamma^2 u_x^2}{c^4}\frac{\partial^2}{\partial t'^2} - \frac{2\gamma^2 u_x}{c^2}\frac{\partial^2}{\partial x'\partial t'} \\ \frac{\partial}{\partial t} &= \frac{\partial x'}{\partial t}\frac{\partial}{\partial x'} + \frac{\partial t'}{\partial t}\frac{\partial}{\partial t'} \\ &= -u_x\gamma\frac{\partial}{\partial x'} + \gamma\frac{\partial}{\partial t'} \\ \Rightarrow \frac{\partial^2}{\partial t^2} &= \frac{\partial}{\partial t}\frac{\partial}{\partial t} \\ \frac{\partial^2}{\partial t^2} &= \left[-u_x\gamma\frac{\partial}{\partial x'} + \gamma\frac{\partial}{\partial t'}\right]\left[-u_x\gamma\frac{\partial}{\partial x'} + \gamma\frac{\partial}{\partial t'}\right] \\ &= u_x^2\gamma^2\frac{\partial^2}{\partial x'^2} + \gamma^2\frac{\partial^2}{\partial t'^2} - 2u_x\gamma^2\frac{\partial^2}{\partial x'\partial t'} \end{aligned} \tag{8.34}$$

Substituting the derivative expressions from Eq. (8.34) into the 1D wave equation, we get

$$
\begin{aligned}
&\left(\frac{\partial^2}{\partial x^2} - \frac{1}{c^2}\frac{\partial^2}{\partial t^2}\right)\Psi = 0 \\
\Rightarrow &\left[\gamma^2\frac{\partial^2}{\partial x'^2} + \frac{\gamma^2 u_x^2}{c^4}\frac{\partial^2}{\partial t'^2} - \frac{2\gamma^2 u_x}{c^2}\frac{\partial^2}{\partial x'\partial t'}\right. \\
&\left. - \frac{1}{c^2}\left(u_x^2\gamma^2\frac{\partial^2}{\partial x'^2} + \gamma^2\frac{\partial^2}{\partial t'^2} - 2u_x\gamma^2\frac{\partial^2}{\partial x'\partial t'}\right)\right]\Psi = 0 \\
\Rightarrow &\left(\frac{\partial^2}{\partial x'^2} - \frac{1}{c^2}\frac{\partial^2}{\partial t'^2}\right)\Psi = 0
\end{aligned}
\tag{8.35}
$$

which clearly shows that the wave equation is invariant under the Lorentz transformation. One can also show that all the individual Maxwell's equations are also invariant under Lorentz transformation. However, it is important to note that although these equations remain invariant under the Lorentz transformation, the individual EM fields do not! For the case of the S' frame moving along the $+x$ direction with respect to S frame (without loss of generality), the transformation rules for the EM fields, the potentials, charge and current densities are given by

$$
\begin{aligned}
E_x' &= E_x \\
E_y' &= \gamma(E_y - u_x B_z) \\
E_z' &= \gamma(E_z + u_x B_y)
\end{aligned}
\tag{8.36}
$$

$$
\begin{aligned}
B_x' &= B_x \\
B_y' &= \gamma\left(B_y + \frac{u_x}{c^2}E_z\right) \\
B_z' &= \gamma\left(B_z - \frac{u_x}{c^2}E_y\right)
\end{aligned}
\tag{8.37}
$$

$$
\begin{aligned}
V' &= \gamma(V - u_x A_x) \\
A_x' &= \gamma\left(A_x - \frac{u_x}{c^2}V\right) \\
A_y' &= A_y \\
A_z' &= A_z
\end{aligned}
\tag{8.38}
$$

$$
\begin{aligned}
\rho' &= \gamma\left(\rho - \frac{u_x}{c^2}J_x\right) \\
J_x' &= \gamma(J_x - u_x\rho) \\
J_y' &= J_y \\
J_z' &= J_z
\end{aligned}
\tag{8.39}
$$

A comparison of Eqs. (8.33) and (8.36) seems to suggest that the electric field transforms like the space coordinate and the magnetic field transforms like the time-coordinate. Hence, the electric and magnetic field are perhaps not two separate entities but form a unified electromagnetic field in the same sense as the idea of a unified 4D space-time instead of separate three dimensions of space and one dimension of time. A comparison of the above 4 equations also shows that the fields transform in a way complimentary to the way in which the vector potential and currents transform. The parallel component of the electric and magnetic field remains invariant, whereas the perpendicular component of the vector potential and current remains invariant. In Eq. (8.39), it is important to note that the electric charge is invariant of coordinate transformations and the charge density undergoes a transformation only due to a change of the volume element.

In the formulation of special relativity, the electric and magnetic fields are not written separately but as the components of a 4 × 4 tensor,

$$F^{\mu\nu} = \begin{pmatrix} 0 & -E_x/c & -E_y/c & -E_z/c \\ E_x/c & 0 & -B_z & B_y \\ E_y/c & B_z & 0 & -B_x \\ E_z/c & -B_y & B_x & 0 \end{pmatrix} \tag{8.40}$$

and it can be shown that this tensor remains invariant under the Lorentz transformation, which are given by simple rules of matrix multiplication. Similarly, the current-charge densities and the scalar-vector potentials are written as 4D vectors

$$J^\alpha = (c\rho, J_x, J_y, J_z) \tag{8.41}$$

$$A^\alpha = \left(V/c, A_x, A_y, A_z\right) \tag{8.42}$$

Hence, the relation between the electromagnetic fields and the potentials is given by

$$F^{\alpha\beta} = \frac{\partial A^\beta}{\partial x_\alpha} - \frac{\partial A^\alpha}{\partial x_\beta} \tag{8.43}$$

where the position vector is given by

$$x_\alpha = (ct, -x, -y, -z) \tag{8.44}$$

In this notation, the four Maxwell's equations can be written in a compact form,

$$\frac{\partial F^{\alpha\beta}}{\partial x^\alpha} = \mu_0 J^\beta$$

where, as per the Einstein summation convention, the presence of the same index α in the numerator and denominator implies a sum over this index.

8.3 LTI Systems and Green's Function

Maxwell's equations can also be seen from the perspective of a system where the charge/current is the input and the electric/magnetic field (or the scalar/vector potential) is the output, as shown in Figure 8.7. A common example is an antenna problem where we have a current carrying wire of a pre-determined shape and current distribution and the task is to find the resulting EM field generated. There are also situations where an EM field can be the input and another EM field the output. This occurs in scattering problems where an externally applied incoming EM field (usually a plane wave, denoted by $\vec{E}_e$ and $\vec{B}_e$ in the figure) encounters an obstacle and produces the scattered EM field as the final output. An important class of such systems are ones which have the properties of *linearity* and *time-invariance*. In order to understand these properties, we need to take recourse to their mathematical formulation.

Consider the system as shown in Figure 8.7 which takes an input $x(t)$ and produces an output $y(t)$. Let $x_1(t)$ and $x_2(t)$ be two different inputs to this system which produce outputs $y_1(t)$ and $y_2(t)$ respectively. This system will be said to be linear, if it satisfies the following two properties,

1. Output corresponding to $\alpha x(t)$ will be $\alpha y(t)$, where α is a non-zero scalar quantity.
2. Output corresponding to a linear combination of $x_1(t)$ and $x_2(t)$ will be the same linear combination of $y_1(t)$ and $y_2(t)$,

$$\begin{aligned} x_3(t) &= \alpha_1 x_1(t) + \alpha_2 x_2(t) \\ \Rightarrow y_3(t) &= \alpha_1 y_1(t) + \alpha_2 y_2(t) \end{aligned}$$

And this system will be said to be time-invariant, if a shift of the input by a certain finite time simply results in a shift of the output by the same finite time without leading to any other change in the output,

$$\begin{aligned} x'(t) &= x(t - t_0) \\ \Rightarrow y'(t) &= y(t - t_0) \end{aligned}$$

A system which has both these properties of linearity (L) and time-invariance (TI) is said to be an *LTI system*. In Figure 8.7, if we neglect the feedback mechanism and assume the material to be non-dispersive, then the Maxwell's equations also obey the

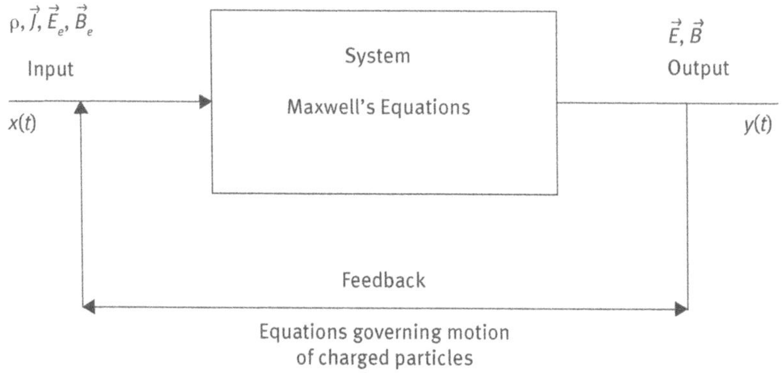

Fig. 8.7: Representation of the Maxwell's equations as a system with feedback.

properties of an LTI system. This property considerably simplifies methods of solving Maxwell's equations in this limiting case (which also has many applications).

The most powerful implication of a system being LTI is that we do not need to solve the full system equations every time we encounter a new input function. All that is required is to compute the output to the *Dirac delta function*, $\delta(t)$, and then the output to any other input can be written using this output known as the *impulse response* and usually denoted by $h(t)$. The Dirac delta function is a very useful tool in physics and denotes a function that is zero for all $t \neq 0$ and has an area of unity when integrated in the range $t \in [-\infty, \infty]$. In order to visualise this better, consider a rectangular pulse which has a non-zero constant value of $1/2T$ within the range $t \in [-T, T]$ (and zero everywhere else). Now, as we let $T \to 0$, the rectangular pulse turns into the Dirac delta function. Any arbitrary input signal, $x(t)$, can be written as an integral over this delta function,

$$x(t) = \int_{-\infty}^{\infty} x\left(t'\right) \delta\left(t - t'\right) dt' \tag{8.45}$$

Now, if the system under consideration is LTI, then its output is simply given by

$$\begin{aligned} y(t) &= \int_{-\infty}^{\infty} x\left(t'\right) h\left(t - t'\right) dt' \\ &= \int_{-\infty}^{\infty} h\left(t'\right) x\left(t - t'\right) dt' \end{aligned} \tag{8.46}$$

which is usually denoted using the symbol $\star$ known as the *convolution*,

$$\begin{aligned} y(t) &= x(t) \star h(t) \\ &= h(t) \star x(t) \end{aligned} \tag{8.47}$$

Let us know use this theory in solving the electromagnetic wave equations, Eqs. (8.26) and (8.27). The equation for the scalar potential is given by

$$\nabla^2 V - \frac{1}{c^2} \frac{\partial^2 V}{\partial t^2} = -\frac{\rho_q}{\epsilon_0} \tag{8.48}$$

which clearly obeys the principle of LTI systems (assuming that the ρ_q is not influenced by the potentials, which is a reasonable assumption in many practical applications). An arbitrary charge distribution can be represented by using Dirac delta functions in the following way,

$$\rho_q(\vec{r}, t) = \int \rho_q\left(\vec{r}', t'\right) \delta\left(\vec{r} - \vec{r}'\right) \delta\left(t - t'\right) d\tau' dt' \tag{8.49}$$

where $d\tau'$ is the volume integral. So now if we can solve Eq. (8.48) for a delta function charge density input, $\delta\left(\vec{r} - \vec{r}'\right) \delta\left(t - t'\right)$, we can write the final solution for an arbitrary charge density as

$$V(\vec{r}, t) = \int \rho_q\left(\vec{r}', t'\right) G\left(\vec{r}, \vec{r}'; t, t'\right) d\tau' dt' \tag{8.50}$$

where $G\left(\vec{r},\vec{r}';t,t'\right)$ is known as the *Green's function* and is a solution to the equation,

$$\left(\nabla^2 - \frac{1}{c^2}\frac{\partial^2}{\partial t^2}\right) G\left(\vec{r},\vec{r}';t,t'\right) = -\frac{1}{\epsilon_0}\delta\left(\vec{r}-\vec{r}'\right)\delta\left(t-t'\right) \tag{8.51}$$

This equation can be solved to obtain

$$G\left(\vec{r},\vec{r}';t,t'\right) = \frac{1}{4\pi\epsilon_0}\frac{\delta\left(t-t'-\left|\vec{r}-\vec{r}'\right|/c\right)}{\left|\vec{r}-\vec{r}'\right|} \tag{8.52}$$

Substituting this in Eq. (8.50), we get

$$\begin{aligned} V\left(\vec{r},t\right) &= \int \rho_q\left(\vec{r}',t'\right) G\left(\vec{r},\vec{r}';t,t'\right) d\tau' dt' \\ &= \frac{1}{4\pi\epsilon_0}\int \frac{\rho_q\left(\vec{r}',t_r\right)}{\left|\vec{r}-\vec{r}'\right|} d\tau' \end{aligned}$$

where $t_r = t - \left|\vec{r}-\vec{r}'\right|/c$ is the retarded time, and which is exactly same as that given by Eq. (8.28). The solution for the vector potential follows the exact same method.

Among several properties of LTI systems, two of the most important ones are *causality* and *stability*. Causality refers to the property that the output at a certain time is effected only input at that particular time or previous times. It might appear that every single system on this planet must be causal, but that is not so in practice. In many problems in signal and image processing, the available input is analysed as a whole after all the data has been collected. In such cases, the output at a certain point can depend on points prior to or later than this point. But in many other problems, mainly those where real-time data analysis is done, causality is an important criteria for the system to be meaningful. And stability is, of course, an important criteria for almost all systems and implies that the output will remain bounded as along as the input is bounded.

For a given LTI system, these two properties of causality and stability can be easily ascertained through properties of the impulse response, $h(t)$. For a system to be causal, the $h(t)$ must be zero for all times $t < 0$. In this case, the convolution integral becomes

$$\begin{aligned} y(t) &= \int_{-\infty}^{\infty} h\left(t'\right) x\left(t-t'\right) dt' \\ y(t) &= \int_{0}^{\infty} h\left(t'\right) x\left(t-t'\right) dt' \end{aligned} \tag{8.53}$$

which obviously implies that the output $y(t)$ at an arbitrary time t_1 is effected by inputs only at times $t \le t_1$. For the system to be stable, for a given bounded input ($|x(t)| \le B_x$ for all $t \in \Re$), we must have $|y(t)| \le B_y$ for all $t \in \Re$, where B is a non-zero finite real number. Using the convolution integral, let us evaluate the modulus of $y(t)$,

$$\begin{aligned}|y(t)| &= \left| \int_{-\infty}^{\infty} x\left(t'\right) h\left(t-t'\right) dt' \right| \\ &\leq \int_{-\infty}^{\infty} \left|x\left(t'\right)\right| \left|h\left(t-t'\right)\right| dt' \qquad \text{Triangle Inequality} \\ &\leq B_X \int_{-\infty}^{\infty} \left|h\left(t-t'\right)\right| dt' \\ \Rightarrow |y(t)| &\leq B_X \int_{-\infty}^{\infty} |h(t)|\, dt\end{aligned}$$

This implies that if the impulse response is absolutely integrable, i. e.,

$$\int_{-\infty}^{\infty} |h(t)|\, dt < \infty$$

then the LTI system will be stable.

8.4 Fourier Transform

The convolution integral given by Eq. (8.46) looks quite simple, but it can be quite tedious to evaluate in many situations. Also, the impulse response of an LTI system, $h(t)$, by itself does not say much about its properties apart from its causality and stability. As it turns out, there is a different way of representing the impulse response which is much easier to use and is also very informative. This alternate representation is known as the *transfer function* , $H(\omega)$, and is obtained by applying what is known as the *Fourier transform* to the impulse response.

In the beginning of this chapter, we studied vector quantities and saw that they can represented by a vector sum of their components along the three coordinate axes. Another way of saying this is that the three coordinate axes *span* the entire 3D space. A similar idea exists in the abstract space of functions too. Consider the set (or space), S_f, consisting of all absolutely integrable functions of a single variable, t,

$$S_f = \left\{ f(t) : \int_{-\infty}^{\infty} |f(t)|\, dt < \infty \right\}$$

It can be shown that the set of functions of the form, $e^{i\omega t}$, spans this set S_f, where $i = \sqrt{-1}$ and $\omega \in \Re$, and the representation of functions, $f(t) \in S_f$, in terms of the functions $e^{i\omega t}$ is given by

$$f(t) = \frac{1}{2\pi} \int_{-\infty}^{\infty} F(\omega)\, e^{i\omega t} d\omega \tag{8.54}$$

where the function, $F(\omega)$, is the Fourier transform of $f(t)$. An important point here is that the functions, $e^{i\omega t}$, also known as a phaser itself does not belong to S_f. Like in the position vector, $\vec{P} = x\hat{x} + y\hat{y} + z\hat{z}$, x is the magnitude of the vector along the $\hat{x}$ direction, similarly $F(\omega)$ is the magnitude of the function along the $e^{i\omega t}$ component for any arbitrary value of ω. The factor $1/2\pi$ in Eq. (8.54) is just a matter of convention. We can also invert Eq. (8.54) to obtain $F(\omega)$ for any given function, $f(t)$. In order to do this, multiply both sides of Eq. (8.54) by $e^{-i\omega' t}$ and integrate with respect to t,

$$\begin{aligned}
\int_{-\infty}^{\infty} f(t)\, e^{-i\omega' t} dt &= \frac{1}{2\pi}\int_{-\infty}^{\infty}\int_{-\infty}^{\infty} F(\omega)\, e^{i\omega t} e^{-i\omega' t} d\omega dt \\
&= \frac{1}{2\pi}\int_{-\infty}^{\infty} F(\omega)\left[\int_{-\infty}^{\infty} e^{i(\omega-\omega')t} dt\right] d\omega \\
&= \frac{1}{2\pi}\int_{-\infty}^{\infty} F(\omega)\left[2\pi\delta\left(\omega-\omega'\right)\right] d\omega \\
&= F\left(\omega'\right) \\
\Rightarrow F(\omega) &= \int_{-\infty}^{\infty} f(t)\, e^{-i\omega t} dt
\end{aligned} \tag{8.55}$$

Now, what is the Fourier transform of the phaser itself? It is clear that $e^{i\omega_0 t}$ is not absolutely integrable and so does not belong to the set S_f. The Fourier transform of this phaser is written using the Dirac delta function and is given by $2\pi\delta(\omega - \omega_0)$. This result was used in the derivation of Eq. (8.55) above.

So how does Fourier transform help in solving for the output of LTI systems? In order to see this, take the Fourier transform of the output, $y(t)$, given by Eq. (8.46),

$$\begin{aligned}
Y(\omega) &= \int_{-\infty}^{\infty} y(t)\, e^{-i\omega t} dt \\
&= \int_{-\infty}^{\infty}\left[\int_{-\infty}^{\infty} x\left(t'\right) h\left(t-t'\right) dt'\right] e^{-i\omega t} dt \\
&= \int_{-\infty}^{\infty}\left[\int_{-\infty}^{\infty} h\left(t-t'\right) e^{-i\omega(t-t')} dt\right] x\left(t'\right) e^{-i\omega t'} dt' \\
&= \int_{-\infty}^{\infty}\left[\int_{-\infty}^{\infty} h(t)\, e^{-i\omega t} dt\right] x\left(t'\right) e^{-i\omega t'} dt' \\
&= \int_{-\infty}^{\infty} H(\omega)\, x\left(t'\right) e^{-i\omega t'} dt' \\
\Rightarrow Y(\omega) &= X(\omega)\, H(\omega)
\end{aligned} \tag{8.56}$$

where the function, $H(\omega)$, is known as the transfer function of the LTI system under consideration. It is clear from Eq. (8.56) that a Fourier transform converts a convolution integral in time domain to a simple multiplication in the frequency domain. And the reverse is also true. A multiplication in the time domain becomes a convolution in the frequency domain. This follows from the fact that the integrals in Eqs. (8.54) and (8.55) are exactly the same except for a sign of the phaser and a scalar factor. So all the mathematical relations which hold while going from time-domain to frequency domain are also valid while going from the frequency domain to the time domain.

So, if we know any arbitrary input-output pair of a given LTI system, in order to find the transfer function, we just need to take the ratio of the Fourier transform of the output and the input. And now this transfer function can be used to compute the output for any other input function. As compared to the convolution formulation in time-domain, the frequency-domain Fourier transform formulation of an LTI system is not only easier to use analytically but also numerically. It is numerically preferable since there are algorithms (e. g., Fast Fourier Transform, FFT) which make numerical computation of the Fourier transform and its inverse really very fast.

In the case of Maxwell's equations, the EM fields are functions of both space and time. The Fourier transform variable corresponding to the time-dimension is frequency, ω, and the variable corresponding to the space-dimension is wave number, $\vec{k} = (k_x, k_y, k_z)$. For the case of plane EM waves in free space, we know that $\omega = ck$, where c is the speed of light. In Eqs. (8.54) and (8.55), the sign of ω in the phaser term, $e^{i\omega t}$, inside the integral can be interchanged without loss of generality. The sign used in a particular problem is only a matter of convention and does not make any difference to the result as long as we are consistent. In this book, for computing Fourier transform of the various field components, we use the phaser, $\exp\left[i\left(\vec{k}\cdot\vec{r}-\omega t\right)\right]$, in Eq. (8.54),

$$\vec{E}\left(\vec{r},t\right) = \frac{1}{2\pi}\int_{-\infty}^{\infty}\vec{\tilde{E}}\left(\vec{k},\omega\right)e^{i\left(\vec{k}\cdot\vec{r}-\omega t\right)}d^3k\,d\omega \tag{8.57}$$

but in some other references, the opposite convention is also used.

Though Fourier transform can be used to obtain the output of any LTI system for any given input (that is absolutely integrable), its usefulness is much more heightened for the case of LTI systems that can be represented as Linear Constant Coefficient Differential Equations (LCCDEs). All LCCDEs are LTI systems, but the reverse is not true. The general form of an LCCDE is given by

$$\sum_{n=0}^{N} a_n \frac{d^n y}{dt^n} = \sum_{m=0}^{M} b_m \frac{d^m x}{dt^m} \tag{8.58}$$

where the coefficients a_n, b_m are constants for $n = 0, 1, 2, 3, \ldots, N$ and $m = 0, 1, 2, 3, \ldots, M$. In order to solve Eq. (8.58), we need find an expression for the Fourier transform of the derivatives of a function. Let us go back to Eq. (8.54) and take a time-derivative on both sides,

$$\begin{aligned}
\frac{df(t)}{dt} &= \frac{1}{2\pi}\frac{d}{dt}\int_{-\infty}^{\infty}F(\omega)\,e^{i\omega t}d\omega \\
&= \frac{1}{2\pi}\int_{-\infty}^{\infty}F(\omega)\left(\frac{d}{dt}e^{i\omega t}\right)d\omega \\
&= \frac{1}{2\pi}\int_{-\infty}^{\infty}i\omega F(\omega)\,e^{i\omega t}d\omega \\
\mathcal{FT}\left[\frac{df(t)}{dt}\right] &= i\omega F(\omega)
\end{aligned}$$

By applying the derivative multiple times in the same manner as shown above, we get,

$$\mathfrak{FT}\left[\frac{d^n f(t)}{dt^n}\right] = (i\omega)^n F(\omega) \tag{8.59}$$

Table 8.1 lists few other important properties of Fourier transform which are very useful in solving differential equations. Substituting this result in Eq. (8.58), we get

$$\sum_{n=0}^{N} a_n (i\omega)^n Y(\omega) = \sum_{m=0}^{M} b_m (i\omega)^m X(\omega)$$
$$\Rightarrow H(\omega) = \frac{Y(\omega)}{X(\omega)} = \frac{\sum_{m=0}^{M} b_m (i\omega)^m}{\sum_{n=0}^{N} a_n (i\omega)^n} \tag{8.60}$$

which is all that we need to solve for the output of this LTI system for any arbitrary input. Given any input, find its Fourier transform by using Eq. (8.55), multiply this resulting function with the transfer function, $H(\omega)$, and then take an inverse Fourier transform of the product using Eq. (8.54).

Maxwell's equations are also linear where the coefficients (permittivity and permeability) are constants in non-dispersive media. But as we have seen in earlier chapters, the more interesting materials are those whose permittivity and/or permeability is dependent on frequency. This is crucial for the existence of plasmonics and many other interesting wave phenomenon in plasmas, which is the main subject of this book. If the coefficients in Eq. (8.58) are time-dependent, one can still take a Fourier transform, but its applicability is quite restricted since now a multiplication in time-domain will become a convolution in frequency-domain making further mathematical evaluation very cumbersome. However, we have still been able to use Fourier transform to solve the Maxwell's equations even for dispersive medium where the permittivity does

Tab. 8.1: Properties of Fourier transforms.

Time-domain function, *h* (*t*)	**Frequency-domain function, *H* (*ω*)**
$h(t) = \frac{1}{2\pi}\int_{-\infty}^{\infty} H(\omega)\, e^{i\omega t} d\omega$	$H(\omega) = \int_{-\infty}^{\infty} h(t)\, e^{-i\omega t} dt$
$h(t - t_0)$	$e^{-i\omega t_0} H(\omega)$
$e^{i\omega_0 t} h(t)$	$H(\omega - \omega_0)$
$h(\alpha t)$	$\frac{1}{\lvert\alpha\rvert} H(\omega/\alpha)$
$H(t)$	$2\pi h(-\omega)$
$d^n h/dt^n$	$(i\omega)^n H(\omega)$
$(-it)^n h(t)$	$d^n H/d\omega^n$
$\int_{-\infty}^{t} h(t')\, dt'$	$H(\omega)/i\omega + \pi H(0)\, \delta(\omega)$
$\delta(t)$	1
$e^{t\omega_0 t}$	$2\pi\delta(\omega - \omega_0)$
sgn (t) [sign function]	$2/i\omega$
$u(t)$ [unit-step function]	$\pi\delta(\omega) + 1/i\omega$
$e^{-at} u(t)$	$1/(a + i\omega)$
$e^{-\lvert a\rvert t}$	$2a/(a^2 + \omega^2)$

depend on frequency. This is a tricky point and needs to be understood clearly. For solving the Maxwell's equations in matter, we usually introduce two new field quantities, $\vec{D}$ and $\vec{H}$, which are linearly depended on $\vec{E}$ and $\vec{B}$ through a multiplication. As mentioned in previous chapters, what is important to note is that this multiplication is valid in the frequency domain and not in the time domain. So, the correct relationship between $\vec{D}$ and $\vec{E}$ in the frequency domain is $\vec{D}(\omega) = \epsilon(\omega)\vec{E}(\omega)$, and so in the time domain it becomes a convolution,

$$\vec{D}(t) = \epsilon(t) \star \vec{E}(\omega) \tag{8.61}$$

Thus, the Maxwell's equations are automatically simplified in the frequency-domain even for dispersive materials. Also, in the linear regime, the solution of an equation at one frequency is independent of the result at other frequencies. If the material has nonlinear behaviour (which is what happens at very high intensities of the EM fields), then this is process does not work. But as long as we are in the linear regime, Fourier transform methods can still be used in solving equations, though a lot more care may have to be taken for the case of time-varying coefficients if they appear as multiplicative terms in the time-domain. There are also many other powerful methods used to solve such equations. Though the representation of the electric displacement vector, $\vec{D}$, is very simple in the frequency-domain and can greatly simplify the process of solving Maxwell's equations, there is a wide range of problems where we need to explicitly solve the time-domain equation. In particular, this is true in problems where the time evolution of the waves need to be studied and the complexity of the problem does not permit mathematical analysis, which is quite common in the study of plasmas and surface plasmons. One could also computationally solve the problem in frequency domain and do an inverse Fourier transform to get the time-domain information, but this method is computationally very inefficient for many reasons. Hence, the time-domain equation, Eq. (8.61), is very useful in computational electromagnetics and forms a core part of the FDTD method as discussed in Chapter 2.

8.5 Linear Stability Analysis of ODEs

While discussing optical tweezers in Chapter 7, it was mentioned that the particle is trapped at a point in space where the gradient forces of the focussed laser beam balances its radiation pressure. The radiation pressure tries to move the particle along with the beam and the gradient forces (equivalent of ponderomotive effect) try to bring the particle towards the point of maximum laser intensity (where the laser beam has the narrowest waist). In the terminology of dynamical systems [30], this point of equilibrium is called a *fixed point* and there are very simple yet powerful tools available to determine whether this fixed point leads to a stable/unstable/neutral kind of equilibrium. In this section, we will discuss some of these methods. The equations of motion of particles in such systems are in general nonlinear and require very complex tools for a complete analysis, but a lot of insights can be gained even through a linear analysis. In order to do this, we first carry out a linearisation of our equations like we did in Chapters 3 and 4 while studying plasma waves. The linearisation of plasma equations for studying wave phenomenon was done about the equilibrium solutions which are usually known to us a priori. Similarly, in the case of dynamical systems or nonlinear equations of particle motion, we first need to find the point of equilibrium about which to do the linearisation and further analysis.

For the sake of simplicity, consider a nonlinear time-independent 2D system governed by the following equations

$$\begin{aligned}\dot{u} &= f(u, v)\\ \dot{v} &= g(u, v)\end{aligned} \tag{8.62}$$

and let (u^*, v^*) be a fixed point of this system. This implies that $f(u^*, v^*) = 0 = g(u^*, v^*)$. In order to linearise the above nonlinear equation, we need to do a Taylor expansion of the functions, f and g, in the neighbourhood of (u^*, v^*), which gives

$$\begin{aligned}f\left(u^* + x_1, v^* + x_2\right) &= f\left(u^*, v^*\right) + x_1 \left.\frac{\partial f}{\partial u}\right|_{(u^*,v^*)} + x_2 \left.\frac{\partial f}{\partial v}\right|_{(u^*,v^*)}\\ &\quad + \mathcal{O}\left(x_1^2, x_2^2, x_1 x_2\right)\\ &= x_1 \left.\frac{\partial f}{\partial u}\right|_{(u^*,v^*)} + x_2 \left.\frac{\partial f}{\partial v}\right|_{(u^*,v^*)} + \mathcal{O}\left(x_1^2, x_2^2, x_1 x_2\right)\\ g\left(u^* + x_1, v^* + x_2\right) &= g\left(u^*, v^*\right) + x_1 \left.\frac{\partial g}{\partial u}\right|_{(u^*,v^*)} + x_2 \left.\frac{\partial g}{\partial v}\right|_{(u^*,v^*)}\\ &\quad + \mathcal{O}\left(x_1^2, x_2^2, x_1 x_2\right)\\ &= x_1 \left.\frac{\partial g}{\partial u}\right|_{(u^*,v^*)} + x_2 \left.\frac{\partial g}{\partial v}\right|_{(u^*,v^*)} + \mathcal{O}\left(x_1^2, x_2^2, x_1 x_2\right)\end{aligned} \tag{8.63}$$

where x_1, x_2 denote small deviations from the fixed point. Substituting this in Eq. (8.62) and neglecting the higher order terms, we get the linearised equations,

$$\begin{aligned}\dot{x}_1 &= x_1 \left.\frac{\partial f}{\partial u}\right|_{(u^*,v^*)} + x_2 \left.\frac{\partial f}{\partial v}\right|_{(u^*,v^*)}\\ \dot{x}_2 &= x_1 \left.\frac{\partial g}{\partial u}\right|_{(u^*,v^*)} + x_2 \left.\frac{\partial g}{\partial v}\right|_{(u^*,v^*)}\end{aligned}$$

which can easily be generalised to a system governed by N first order linear ODEs with constant coefficients,

$$\frac{dx_1}{dt} = a_{11}x_1 + a_{12}x_2 + \cdots + a_{1N}x_N$$

$$\frac{dx_2}{dt} = a_{21}x_1 + a_{22}x_2 + \cdots + a_{2N}x_N$$
$$\vdots$$
$$\frac{dx_N}{dt} = a_{N1}x_1 + a_{N2}x_2 + \cdots + a_{NN}x_N \tag{8.64}$$

where $a_{ij} \in \Re$, obtained after linearisation of the nonlinear equations as shown above, which can also be written in the more useful matrix notation,

$$\frac{d}{dt}\begin{pmatrix} x_1 \\ x_2 \\ \vdots \\ x_N \end{pmatrix} = \begin{pmatrix} a_{11} & a_{12} & \cdots & a_{1N} \\ a_{21} & a_{22} & \cdots & a_{2N} \\ \vdots & \vdots & \vdots & \vdots \\ a_{N1} & a_{N2} & \cdots & a_{NN} \end{pmatrix}\begin{pmatrix} x_1 \\ x_2 \\ \vdots \\ x_N \end{pmatrix} \tag{8.65}$$
$$\dot{\vec{x}} = \overleftrightarrow{A}\vec{x}$$

where $\vec{x}$ is a vector representation of the $N \times 1$ column vector consisting of the N-coordinates, $\{x_1, x_2, \ldots, x_N\}$ and $\overleftrightarrow{A}$ is the $N \times N$ matrix containing the coefficients of the ODE (also known as the *Jacobian matrix*). In the case of particle traps discussed in Chapter 7, the dynamical equations are also time-dependent in addition to being nonlinear. In such cases, the time-dependence is removed by taking an average over the high frequency components (ponderomotive theory) and then the resulting time-independent equations are analysed using the method described in this chapter.

The fixed point corresponding to Eq. (8.65) is obtained by equating the RHS to zero, since at the fixed point, the dynamical variables must be at equilibrium ($\dot{x}_1 = 0 = \dot{x}_2 = \cdots = \dot{x}_N$). Obviously, all equations of the form Eq. (8.65) have a fixed point at $x_1 = 0 = x_2 = \cdots = x_N$, which automatically follows from the linearisation process adopted as shown above.

The next step is to analyse the stability of the fixed point corresponding to Eq. (8.65). Now, as discussed earlier in this chapter in the context of divergence and curls, there are essentially three ways in which a point near the origin can behave. Broadly speaking, a particle starting from these neighbouring initial conditions can either move away from the origin (*unstable* fixed point), or move towards the origin (*stable* fixed point) or go around the origin in circles (*neutral* fixed point). There is also a fourth possibility that needs to be taken into account. Since we are primarily interested in systems with multiple dimensions, it is quite possible for the equations to be *stable* along one dimension and *unstable* along another dimension. In this case, what might happen is that if we start from a point away from the origin, the particle may initially approach the origin along one/more of the dimensions (giving a semblance of stability) but may start diverging along other dimensions after a certain point of closest approach. Fixed points which show such behaviour are called *saddle nodes* or *saddle points*. As discussed earlier, in particle traps, these saddle nodes play a very important role and prevent confinement of charged particles through static

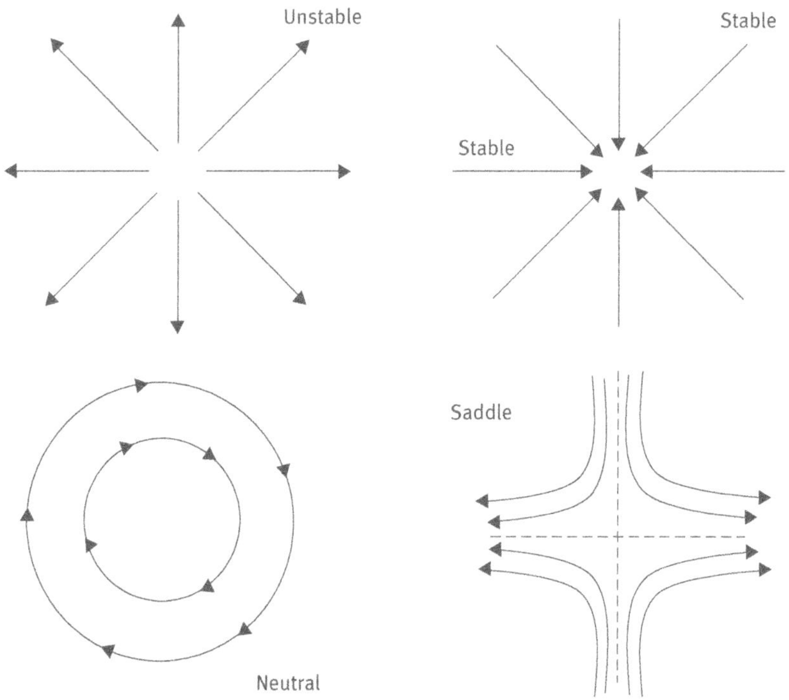

Fig. 8.8: Various kinds of stability for a fixed point.

electric fields (through Earnshaw's theorem). These various possibilities for a fixed point are depicted in Figure 8.8 for the simple case of a two dimensional system. It is important to note that the stable/unstable fixed points can have much more complex flows around them than what is depicted in Figure 8.8. The particle does not always move along straight lines away from or towards these fixed points. There are certain dimensions along which the particle moves much faster than along other dimensions, which result in curved trajectories.

The saddle point depicted in Figure 8.8 is a particular case which shows that the particle trajectory initially moves along the vertical axis towards the fixed point but then starts diverging towards the horizontal axis asymptotically, i. e., as $t \to \infty$. In this case, the vertical axis is called the *unstable manifold* and the horizontal axis is called the *stable manifold*. This is also just a depiction and the actual stable/unstable manifolds can be along any dimension(s). A system in general has several fixed points which may be stable or unstable or neutral or saddle. In the linear stability analysis, the properties of each fixed point is independent of the properties of other fixed points. But in the nonlinear regime, these fixed points can interact with other quite intricately. The stable manifold of one fixed point can feed into the unstable manifold of another fixed point leading to interesting transport properties in the case of fluid mechanics.

It is also possible for these manifolds to fold onto themselves as we move away from the fixed point leading to chaotic properties of particle trajectories.

When we talk of the stability of a fixed point, we usually refer to the behaviour of trajectories starting from neighbouring points for all points in time. It is also possible for a trajectory to be initially diverging away from the fixed point, but approach it after a very long time. This can happen in nonlinear systems. A fixed point is said to be an *attractor* if all trajectories that start in a neighbourhood of this fixed point approach it as $t \to \infty$, irrespective of how the trajectory behaves at transient times. Alternatively, a fixed point is said to be a repeller if all trajectories that start in its neighbourhood approach infinity as $t \to \infty$. If all trajectories starting from a point close to the fixed point stay within its small neighbourhood for all times, then the fixed point is called *Lyapunov stable*. Now, if a point is Lyapunov stable, it does not necessarily mean that the trajectories close to it will converge to it as $t \to \infty$. In fact, *neutrally stable* fixed points are those which are Lyapunov stable but are not attracting. In the absence of friction and other modes of dissipation, many problems in science and engineering have the dynamical structure of an oscillator with fixed points which are neutrally stable. It is also possible for a fixed point to be attracting without being Lyapunov stable. The dynamics of these points needs to be analysed using nonlinear methods, since the various notions of linear stability analysis do not apply here. So, linearised equations will not show this behaviour and the full nonlinear equation needs to be solved to evaluate such behaviour. If a fixed point is both an attractor and is Lyapunov stable, then it is called a stable fixed point as described above. And, if a fixed point is neither attracting nor Lyapunov stable, then it is either unstable or is a saddle node.

Now, let us go back to Eq. (8.65) and discuss ways of mathematically finding out the stability property of the fixed point at origin. If our system is one dimensional, then its linearised dynamical equation is simply given by

$$\dot{x} = \lambda x$$

where the fixed point has been translated to be at the origin and λ is some scalar constant. This equation is simple to solve and gives $x = x_0 e^{\lambda t}$ as the solution. It can be clearly seen that if $\lambda > 0$ is positive, then the fixed point at origin is unstable and if $\lambda < 0$, then it is stable. For the trivial case of $\lambda = 0$, all points on the real line are fixed points and no movement happens at all. In the simple case of 1D motion, there will obviously be no neutral or saddle points. To see these points we need to go to two or higher dimensions. Though the solution of the 1D equation is simplistic, it provides a template for solving the dynamical equations in the case of higher dimensions also. For the N-D system depicted in Eq. (8.65), we look for solutions of the form, $\vec{x} = e^{\lambda t}\vec{u}$, where $\vec{u}$ is a $N \times 1$ column vector consisting of the components along the N-coordinate axes. Substituting this form of the solution in Eq. (8.65), we get,

$$\lambda e^{\lambda t}\vec{u} = \hat{A} e^{\lambda t}\vec{u}$$

$$\Rightarrow \overleftrightarrow{A}\vec{u} = \lambda\vec{u}$$
$$\Rightarrow \left(\overleftrightarrow{A} - \lambda\overleftrightarrow{I}\right)\vec{u} = 0 \qquad (8.66)$$

which is known as the *characteristic equation*, where $\overleftrightarrow{I}$ is the identity matrix.

We clearly see from Eq. (8.66) that λ is the eigenvalue of the matrix $\overleftrightarrow{A}$ and $\vec{u}$ is the eigenvector. An $N \times N$ matrix usually has N eigenvalues and N eigenvectors. Some of these eigenvalues which may be positive, negative or zero. In general, the eigenvalues can also be complex. For the fixed point to be stable, all the corresponding eigenvalues must have negative real part and for the fixed point to be unstable, all the eigenvalues must have positive real part. If the real part of some of the eigenvalues are positive and some negative, then the fixed point is a saddle node and the eigenvectors corresponding to the eigenvalues with positive real part are the unstable manifolds and eigenvectors corresponding to the eigenvalues with negative real part are the stable manifolds. If the eigenvalues are purely imaginary, then the fixed point is neutrally stable.

The stable and unstables nodes depicted in Figure 8.8 are a special case when the fixed point has all eigenvalues equal but N distinct eigenvectors. In this case, all trajectories approach the fixed point at the same rate irrespective of the direction of approach. Such a fixed point is called a *star node*. It is also possible for the matrix $\overleftrightarrow{A}$ to have all N eigenvalues equal and only one distinct eigenvector. In this case, all trajectories become parallel to this eigenvector direction as $t \to \infty$. Such a fixed point is called a *degenerate node*. If all the eigenvalues are equal to zero, then all points in the plane are fixed points.

Let us now take a 2D system and see how these various properties depend on the choice of coefficients contained in the matrix $\overleftrightarrow{A}$,

$$\overleftrightarrow{A} = \begin{pmatrix} a_{11} & a_{12} \\ a_{21} & a_{22} \end{pmatrix}$$

The solutions of the corresponding characteristic equation are given by evaluating the determinant of $|\overleftrightarrow{A} - \lambda\overleftrightarrow{I}|$ and equating it to zero,

$$\left|\overleftrightarrow{A} - \lambda\overleftrightarrow{I}\right| = 0$$
$$\Rightarrow \det\begin{pmatrix} a_{11}-\lambda & a_{12} \\ a_{21} & a_{22}-\lambda \end{pmatrix} = 0$$
$$(a_{11}-\lambda)(a_{22}-\lambda) - a_{21}a_{12} = 0$$
$$\Rightarrow \lambda^2 - T_A\lambda + \Delta_A = 0 \qquad (8.67)$$

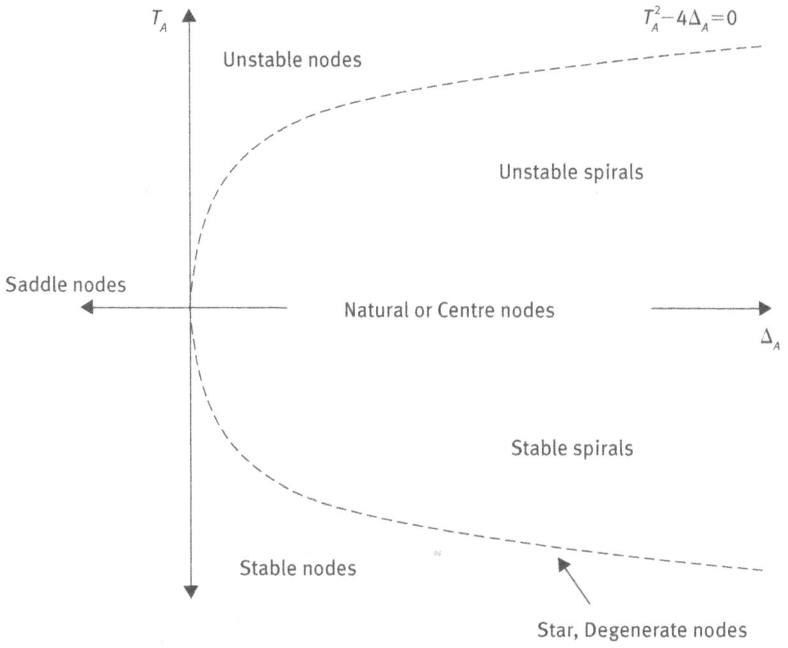

Fig. 8.9: Pictorial representation of stability of a 2D linear system.

where $T_A = a_{11} + a_{22} = \text{Trace}\,(\overleftrightarrow{A})$ and $\Delta_A == a_{11}a_{22} - a_{21}a_{12} = \det\,(\overleftrightarrow{A})$. This equation has two solutions given by

$$\lambda_1 = \frac{1}{2}\left(T_A + \sqrt{T_A^2 - 4\Delta_A}\right)$$

$$\lambda_2 = \frac{1}{2}\left(T_A - \sqrt{T_A^2 - 4\Delta_A}\right)$$

We can draw the following conclusions from an analysis of the above expressions of the two eigenvalues (also depicted pictorially in Figure 8.9),

1. Both the eigenvalues are real when $T_A^2 - 4\Delta_A > 0$ and both are complex if this quantity is negative.
2. If $T_A^2 - 4\Delta_A > 0$, $T_A > 0$ and $\Delta_A > 0$, then both the eigenvalues are also positive, since then $T_A > \sqrt{T_A^2 - 4\Delta_A}$, and the fixed point is unstable.
3. If $T_A^2 - 4\Delta_A > 0$, $T_A < 0$ and $\Delta_A > 0$, then both the eigenvalues are also negative, and the fixed point is stable.
4. If $\Delta_A < 0$, then it automatically follows that $T_A^2 - 4\Delta_A > 0$, and one eigenvalue is positive and one is negative, which implies that the fixed point is a saddle node.
5. If $T_A^2 - 4\Delta_A < 0$ and $T_A \neq 0$, then the real part of both eigenvalues have the same sign as T_A. In this case, the fixed point is a stable spiral (attractor) when $T_A < 0$ and is an unstable spiral (repeller) when $T_A > 0$.

6. If $T_A^2 - 4\Delta_A = 0$, then both the eigenvalues are equal and we either have a degenerate node or a star node, depending on whether we have only one eigenvector or two distinct eigenvectors respectively.
7. If the real part of both eigenvalues is non-zero, then the fixed point is called *hyperbolic*. In this case, the stability property obtained from linear analysis is robust to addition of nonlinear terms in the Taylor expansion of the full nonlinear equation (see Eq. (8.63)). The Hartman-Grobman theorem states that the nonlinear stability property of a hyperbolic fixed point is quantitatively same (phase portraits are topological equivalent) as that obtained from the linearised equation. Two phase portraits (particle trajectories in the N-D space of all coordinates) are said to be topologically equivalent if one is a distorted version of another, i. e., one phase portrait can be obtained by bending and warping of the other (ripping open of closed curves and closing of open lines is not allowed). Such hyperbolic fixed points are also said to be *structurally stable*.
8. For the eigenvalues to be purely imaginary, we must have $T_A = 0$, in which case the fixed point is neutrally stable (centre node). In this case, one of the imaginary eigenvalues has a positive sign and one has a negative sign, implying that one of the eigenvectors rotates clockwise about the fixed point and the other eigenvector rotates anti-clockwise. Centre nodes are not structurally stable since even a small amount of damping in the equation can convert it into a stable fixed point (attractor).

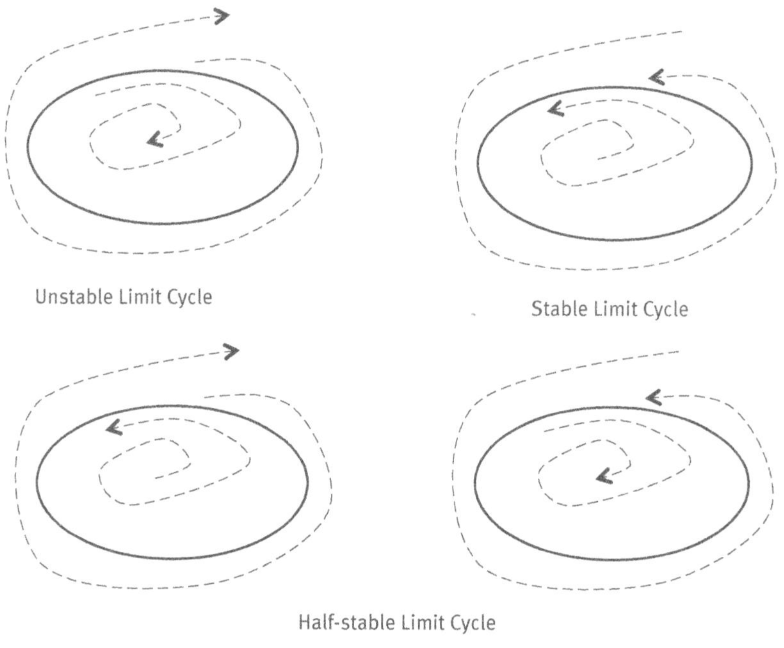

Unstable Limit Cycle

Stable Limit Cycle

Half-stable Limit Cycle

Fig. 8.10: Limit Cycles of a nonlinear dynamical system.

When we consider nonlinear terms in the equation governing a dynamical system, the concept of a fixed point can be generalised to the concept of an isolated closed trajectory called *limit cycle*. Here, the term *isolated* is very important. In the case of a linearised system also, the fixed point can be a centre node, in which case, all neighbouring trajectories are closed curves. These cannot be said to be limit cycles. As shown in Figure 8.10, for a closed trajectory to be a limit cycle, no other trajectory in its neighbourhood must be closed and must either converge towards it (stable limit cycle) or diverge away from it (unstable limit cycle). There are also some rare cases of limit cycles which are *half-stable*, as depicted in Figure 8.10. It is important to note that limit cycles are purely nonlinear phenomenon and cannot occur in linearised systems. Limit cycles are very important in both physical and biological systems. Most of the life systems and natural systems like heartbeat, hormone secretions, seasonal changes of weather, etc. have periodic cycles which essentially behave like stable limit cycles. This is because, for the living organisms to survive, small perturbations about the limit cycles must spiral towards it. In the case of bridges and aeroplane wings, the corresponding rhythmic motion can act as an unstable limit cycle leading to dangerous oscillations of very high amplitude.

A simple case of a 2D system which has a limit cycle is the following,

$$\begin{aligned}\dot{\rho} &= \rho\left(1-\rho^2\right)\\ \dot{\phi} &= 1\end{aligned} \tag{8.68}$$

where ρ, ϕ are the cylindrical polar coordinates, i. e., $\rho = \sqrt{x^2+y^2}$, $\phi = \tan^{-1}\left(y/x\right)$ and x, y are the cartesian coordinates. The above equation clearly has two fixed points, $\rho^* = 0$ (origin) and $\rho^* = 1$ (circle with unity radius). In order to analyse the stability of these two fixed points, we need to linearise Eq. (8.68) about these two points separately. Linearising this equation about the fixed point $\rho^* = 0$, we get

$$\dot{\rho}_1 = \rho_1 + \mathcal{O}\left(\rho_1^2\right)$$

where $\rho_1 = \rho - \rho^*$, which clearly shows that the fixed point at $\rho^* = 0$ is unstable. Linearising Eq. (8.68) about the other fixed point, $\rho^* = 1$ gives

$$\begin{aligned}\dot{\rho}_1 &= (1+\rho_1)\left(1-(1+\rho_1)^2\right)\\ &= (1+\rho_1)\left(-2\rho_1-\rho_1^2\right)\\ &= -2\rho_1 + \mathcal{O}\left(\rho_1^2\right)\end{aligned}$$

which clearly shows that the fixed point at $\rho^* = 1$ is stable. In the 2D space of $x-y$ coordinates, the fixed point at $\rho^* = 1$ is given by the circle, $x^2+y^2 = 1$, and hence corresponds to a limit cycle. The other fixed point at $\rho^* = 0$ remains a fixed point at $x = 0 = y$.

Evaluating whether a given arbitrary system of nonlinear differential equations contains a limit cycle and, if it exists, identifying its trajectory is not very easy. However, the Poincare-Bendixson theorem states that if a trajectory corresponding to a set of equations, $\dot{x}_i = f_i(x_1, x_2, \ldots, x_N)$, for $i = 1, 2, 3, \ldots, N$, remains bounded within a certain closed planar region (starts within this planar region and always stays within it), then this trajectory is either a limit cycle or it tends towards it asymptotically, provided that the closed region does not contain any fixed point and the vector field corresponding to the functions, $f_i(x_1, x_2, \ldots, x_N)$, is continuously differentiable. Though this theorem is very powerful, its applicability is very limited since it requires the trajectory to remain bounded on a planar region, which is usually the case only in 2D systems and in certain very specific higher dimensional systems. This theorem also shows that the dynamics of 2D systems are quite ordered and very limited in possibilities due to the presence of fixed points and limit cycles which give some sort of a rigid structure to the phase portrait. However, these restrictions do not apply to 3D and higher dimensional systems where the trajectories need not be restricted to a plane. This is also one reason why trajectories cannot be *chaotic* in 2D systems and a minimum of three dimensions are required for a system to be chaotic.

8.6 Hamiltonian Formulation of Charged Particle Dynamics

A sub-class of general dynamical systems are those systems whose governing equations of motion can be written in the form

$$\begin{aligned} \dot{q}_i &= \frac{\partial H}{\partial p_i} \\ \dot{p}_i &= -\frac{\partial H}{\partial q_i} \end{aligned} \tag{8.69}$$

where $\{q_i\}$ are the configuration coordinate, $\{p_i\}$ are the *canonical* momenta coordinate and the function, $H(q_1, q_2, \ldots, q_N; p_1, p_2, \ldots, p_N; t)$, is known as the *Hamiltonian*. In usual Newtonian mechanics formulation, the momentum is given by mass times the configuration coordinate. But that is not necessarily the case in Hamiltonian formulation. Here the $q - p$ coordinates are defined such that the resulting equations of motion can be written in the form given by Eq. (8.69). There are many powerful mathematical techniques that can be used in the study of dynamical systems if the corresponding equations can be written in the form of a Hamiltonian. This Hamiltonian formulation is also very useful in the study of quantum mechanics (in both relativistic and non-relativistic cases) since the Hamiltonian function of a system is often the starting point for quantization of the corresponding variables.

For a broad class of systems, the Hamiltonian is simply given by the sum of its kinetic energy and its potential energy. A simple example is a particle moving in a 1D region with a electrostatic potential, $\Phi_{es}(x)$. Here, the equation of motion of the particle is

$$\ddot{x} = -\frac{Q}{M}\frac{\partial \Phi_{es}}{\partial x}$$
$$\Rightarrow \dot{x} = v = p/M$$
$$\text{and } \dot{p} = -Q\frac{\partial \Phi_{es}}{\partial x}$$
$$\Rightarrow \dot{x} = \frac{\partial}{\partial p}\left(\frac{p^2}{2M} + Q\Phi_{es}\right)$$
$$\text{and } \dot{p} = -\frac{\partial}{\partial x}\left(\frac{p^2}{2M} + Q\Phi_{es}\right)$$

which is clearly the Hamiltonian formulation, where the Hamiltonian function is given by

$$H = \frac{p^2}{2M} + Q\Phi_{es} \tag{8.70}$$

which is equal to the total energy of the system and a conserved quantity. This is always the case in time-independent (or *autonomous*) systems. But in systems which are time-dependent or non-autonomous (forcing function is explicitly dependent on time), the Hamiltonian is not a conserved quantity and its time evolution is given by

$$\frac{dH}{dt} = \frac{\partial H}{\partial t} + \sum_i \left\{\frac{dq_i}{dt}\frac{\partial H}{\partial q_i} + \frac{dp_i}{dt}\frac{\partial H}{\partial p_i}\right\}$$
$$= \frac{\partial H}{\partial t} + \sum_i \left\{\frac{\partial H}{\partial p_i}\frac{\partial H}{\partial q_i} - \frac{\partial H}{\partial q_i}\frac{\partial H}{\partial p_i}\right\}$$
$$\Rightarrow \frac{dH}{dt} = \frac{\partial H}{\partial t} \tag{8.71}$$

Equation (8.70) gives the Hamiltonian function for a particle moving in under the influence of an electrostatic potential. Now let us see if particle motion under an arbitrary spatially and temporally varying electromagnetic field can also be written in the form of a Hamiltonian. The Lorentz force equation is given by,

$$\frac{d\vec{v}}{dt} = \frac{Q}{M}\left[\vec{E} + \vec{v} \times \vec{B}\right]$$
$$\Rightarrow \frac{dv_x}{dt} = \frac{Q}{M}\left[E_x + v_y B_z - v_z B_y\right]$$
$$\Rightarrow M\frac{dv_x}{dt} = -Q\frac{\partial \Phi}{\partial x} - Q\frac{\partial A_x}{\partial t} + Qv_y\left(\frac{\partial A_y}{\partial x} - \frac{\partial A_x}{\partial y}\right) + Qv_z\left(\frac{\partial A_z}{\partial x} - \frac{\partial A_x}{\partial z}\right)$$
$$\Rightarrow M\frac{dv_x}{dt} + Q\frac{\partial A_x}{\partial t} + Qv_x\frac{\partial A_x}{\partial x} + Qv_y\frac{\partial A_x}{\partial y} + Qv_z\frac{\partial A_x}{\partial z}$$
$$= Qv_x\frac{\partial A_x}{\partial x} + Qv_y\frac{\partial A_y}{\partial x} + Qv_z\frac{\partial A_z}{\partial x} - Q\frac{\partial \Phi}{\partial x}$$

$$\Rightarrow M\frac{dv_x}{dt} + Q\frac{dA_x}{dt}$$
$$= \frac{Q}{M}(p_x - QA_x)\frac{\partial A_x}{\partial x} + \frac{Q}{M}(p_y - QA_y)\frac{\partial A_y}{\partial x}$$
$$+ \frac{Q}{M}(p_z - QA_z)\frac{\partial A_z}{\partial x} - Q\frac{\partial \Phi}{\partial x}$$
$$\Rightarrow \frac{dp_x}{dt} = -\frac{\partial H_{EM}}{\partial x}$$

where $p_x = Mv_x + QA_x$, $\vec{A}(\vec{r}, t)$ is the vector potential (see Eq. (8.23)), $\Phi(\vec{r}, t)$ is the scalar potential (see Eq. (8.24)) and the Hamiltonian is given by

$$H_{EM} = \frac{1}{2M}\left(\vec{p} - Q\vec{A}(\vec{r}, t)\right)^2 + Q\Phi(\vec{r}, t)$$
$$= \frac{1}{2M}(p_x - QA_x(\vec{r}, t))^2 + \frac{1}{2M}(p_y - QA_y(\vec{r}, t))^2$$
$$+ \frac{1}{2M}(p_z - QA_z(\vec{r}, t))^2 + Q\Phi(\vec{r}, t) \tag{8.72}$$

The equation for the configuration variable, $\vec{r}$, clearly satisfies,

$$\frac{d\vec{r}}{\partial t} = \vec{v}$$
$$= \frac{1}{M}\left(\vec{p} - Q\vec{A}(\vec{r}, t)\right) = \frac{\partial H}{\partial p}$$

Thus, particle motion in an arbitrary electromagnetic field can also be written in the form of a Hamiltonian! However, the Hamiltonian function is a little counter-intituitive. Using the usual velocity variable (instead of the canonical momentum), it becomes

$$H_{EM} = \frac{1}{2}Mv^2 + Q\Phi(\vec{r}, t) \tag{8.73}$$

which does not have any trace of magnetic field, or the scalar potential, in it. Hence, in some sense the role of magnetic field is not to change the Hamiltonian function itself, but merely to introduce a new canonical momentum. For the special case of magnetostatics, $\Phi = 0$ and $\partial\vec{A}/\partial t = 0$, the Hamiltonian given by Eq. (8.72) becomes time independent and is a conserved quantity (see Eq. (8.71)). Equation (8.73) implies that in this case, the Hamiltonian is equal to the usual kinetic energy of the particle which must then be a constant of motion, which is a well known fact (the magnetic field only changes the direction of a particle and not its speed or kinetic energy).

The Hamiltonian given by Eq. (8.72) can be used to directly write the Schrodinger equation for the quantum wave function, $\Psi(\vec{r}, t)$, of the particle,

$$i\hbar\frac{\partial\Psi(\vec{r}, t)}{\partial t} = H\Psi(\vec{r}, t)$$
$$\Rightarrow i\hbar\frac{\partial\Psi(\vec{r}, t)}{\partial t} = \frac{1}{2M}\left(-i\hbar\vec{\nabla} - Q\vec{A}(\vec{r}, t)\right)^2\Psi(\vec{r}, t) + Q\Phi(\vec{r}, t)\Psi(\vec{r}, t)$$

where we have written $\vec{p} = i\hbar\vec{\nabla}$ following the standard rules of non-relativistic quantum mechanics. Quantum mechanics is integral to the development of plasmonics since the term *plasmon* actually stands for quantized plasma oscillations (linear waves studied in Chapter 3 and 4). Surface waves travelling on the metal-dielectric interface are called surface plasmons, since in the frequency range of their existence, a metal essentially behaves like a plasma. Strictly speaking, the term surface plasmon applies to the quantized version of this surface waves and the classical continuous waves should be called surface plasmon polartion waves.

A very important property of Hamiltonian systems is that they are *volume preserving* (in phase space), and this follows from the fact that the Hamiltonian vector field is divergence free (Liouville's theorem). In order to see this, consider a $2N$ dimensional column vector, $\vec{r}$, composed of all the configuration and momentum coordinates,

$$\vec{r} = (q_1, q_2, \ldots, q_N, p_1, p_2, \ldots, p_N)^T$$

The Hamiltonian equations of motion can now be written as

$$\frac{d\vec{r}}{\partial t} = \overleftrightarrow{J} \cdot \left(\vec{\nabla}_r H\right) \tag{8.74}$$

where

$$\overleftrightarrow{J} = \begin{pmatrix} 0 & \overleftrightarrow{I}_N \\ -\overleftrightarrow{I}_N & 0 \end{pmatrix}$$

and

$$\vec{\nabla}_r H = \left(\frac{\partial H}{\partial q_1}, \frac{\partial H}{\partial q_2}, \ldots, \frac{\partial H}{\partial q_N}, \frac{\partial H}{\partial p_1}, \frac{\partial H}{\partial p_2}, \ldots, \frac{\partial H}{\partial p_N}\right)^T$$

where $\overleftrightarrow{I}_N$ is an $N \times N$ identity matrix. The RHS of Eq. (8.74), $\overleftrightarrow{J} \cdot \left(\vec{\nabla}_r H\right)$, constitutes the vector field corresponding the motion of particles in $2N$ dimensional phase space and can be easily shown to be divergence free,

$$\begin{aligned} \vec{\nabla}_r \cdot \left(\overleftrightarrow{J} \cdot \left(\vec{\nabla}_r H\right)\right) &= \sum_i \left\{\frac{\partial^2 H}{\partial q_i \partial p_i} - \frac{\partial^2 H}{\partial p_i \partial q_i}\right\} \\ &= 0 \end{aligned}$$

Now consider a set of particles whose trajectories begin to evolve starting from a contiguous volume in $2N$ dimensional phase space, τ, and bounded by a $2N - 1$ dimensional surface, S. The rate of change of this volume is given by

$$\begin{aligned} \frac{d}{dt}\int d\tau &= \int \frac{d\vec{r}}{dt} \cdot d\vec{S} \\ &= \int \left(\overleftrightarrow{J} \cdot \left(\vec{\nabla} H\right)\right) \cdot d\vec{S} \end{aligned}$$

$$\frac{d}{dt}\int d\tau = \int \vec{\nabla}\cdot\left(\hat{J}\cdot\left(\vec{\nabla}H\right)\right)\tau$$
$$= 0$$

where we have used the divergence theorem given by Eq. (8.19). What this implies is that, for a system governed by Hamiltonian dynamics, if a certain collection of particles starts from a contiguous region of space with volume τ and reaches another region of space with volume, τ', after some time, then we must have $\tau' = \tau$. Thus, if a certain volume is being stretched in one direction due to Hamiltonian evolution, then it must be squeezed in another direction so that the total volume remains constant. For this reason, the Hamiltonian formulation cannot be used to analyse dissipative systems, which surely do not preserve phase space volume. This also has serious implications for the stability analysis of dynamical systems. In the phase portrait shown in Figure 8.8, it can be seen that if we take particles in a certain region of space around the fixed point (which is at the origin in this case), then the area of this collection of particles will decrease or increase if the fixed point is unstable or stable, respectively. However, this would violate the volume/area preserving property of Hamiltonian dynamics. Hence, if a system is governed by Hamiltonian dynamics, none of its fixed points can be stable or unstable, but instead must be saddle nodes or centre nodes. This is indeed the case in Paul traps discussed in Section 7.3, where the particle dynamics obeys the Hamiltonian formulation in the absence of dissipation. By Earnshaw's theorem, an electrostatic vector field cannot have local minima or maxima (stable or unstable fixed points) and all the corresponding fixed points are saddle points. This is why we need a time-periodic field, which creates a centre node in phase space around which the particles can move and thereby be confined for long times.

9 Numerical Methods for Electromagnetics

In this chapter, we will discuss some of the basic numerical methods that can be used to perform computer simulations of Maxwell's equations in various conditions. Computational electromagnetics is a very broad field with too many different methods available for different situations. But we will only give a brief introduction to this topic in this book and encourage the reader to refer to other sources [31] for more details.

9.1 Laplace Equation

The simplest equation in electromagnetics is the Laplace equation, which says that in the absence of charges, currents and time varying fields, the electric potential is a solution of $\nabla^2 \Phi = 0$, where ∇^2 is known as the Laplacian operator. The Laplace equation is important not only in electrostatics, but also in thermodynamics and many other areas of science and engineering. We will discuss the methods of solving this equation in 2D for the sake of simplicity, but the method can be easily generalised to any higher dimensions. In 2D cartesian coordinates, the Laplace equation can be written as

$$\frac{\partial^2 \Phi}{\partial x^2} + \frac{\partial^2 \Phi}{\partial y^2} = 0 \tag{9.1}$$

The biggest difference between analytical and computational solutions is that a computer cannot deal with continuous variables and all quantities are stored using discrete variables. So, the first step is to discretize our space so that it can be used in a computer program. Instead of a continuous x coordinate, we now have a discrete set of points $\{(x_i = i\Delta_x, y_j = j\Delta_y)\}$, where Δ_x and Δ_y are the separations between two neighbouring points along the x- and y-axis (see. Figure 9.1). The next step is to write the derivatives in terms of these discrete variables. In order to do this, we need to go back to the basic definition of a derivative. From the first principles of calculus, we know that

$$\begin{aligned} \frac{df(x)}{dx} &= \lim_{\Delta_s \to 0} \frac{f(x+\Delta_x) - f(x)}{\Delta_x} \\ &= \lim_{\Delta_s \to 0} \frac{f(x) - f(x-\Delta_x)}{\Delta_x} \end{aligned}$$

As mentioned above, the primary feature of a computer is that in its representations, Δ_x cannot go to zero and must be a finite quantity. This obviously introduces certain errors, which can be reduced through careful means. Therefore, for a computer, the discretized derivative is given by

$$\begin{aligned} \frac{df(x)}{dx} &\approx \frac{f(x+\Delta_x) - f(x)}{\Delta_x} \\ &\text{OR, } \frac{f(x) - f(x-\Delta_x)}{\Delta_x} \end{aligned} \tag{9.2}$$

https://doi.org/10.1515/9783110570038-149

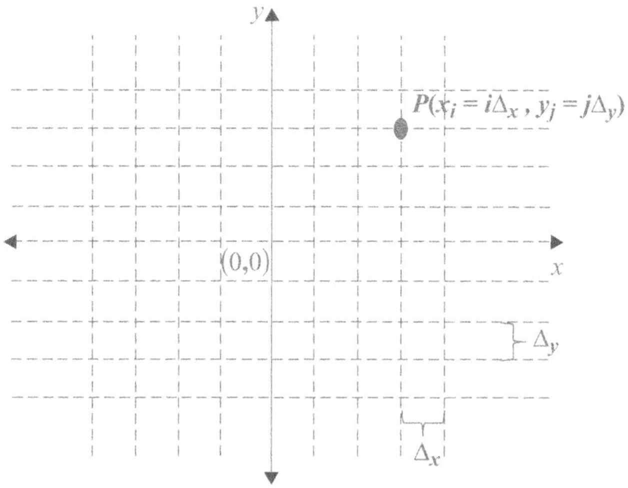

Fig. 9.1: Discretized 2D grid.

In the continuous domain, it does not really matter which of the above two forms we choose since both converge to the same derivative as $\Delta_x \to 0$. However, that is not the case in the discretised domain and one needs to be more careful. Often the choice between these two forms is also a matter of convenience as we will see in the case of Laplace equation. The first form in the above equation is called the *forward propagation* method and the second form is called the *backward propagation* method.

From the first principles, the second derivative can be written as,

$$\begin{aligned}\frac{d^2 f(x)}{dx^2} &= \lim_{\Delta_s \to 0} \frac{f'(x+\Delta_x) - f'(x)}{\Delta_x} \\ &= \lim_{\Delta_s \to 0} \frac{f(x+\Delta_x) - f(x)}{\Delta_x^2} - \frac{f(x) - f(x-\Delta_x)}{\Delta_x^2} \\ &= \lim_{\Delta_s \to 0} \frac{f(x+\Delta_x) - 2f(x) + f(x-\Delta_x)}{\Delta_x^2}\end{aligned}$$

where the prime, f', refers to df/dx, and we have used forward propagation to evaluate the derivative in the 1st step and backward propagation in the 2nd step. Thus, the discretized version of the 2nd derivative is given by

$$\frac{d^2 f}{dx^2} \approx \frac{f(x+\Delta_x) - 2f(x) + f(x-\Delta_x)}{\Delta_x^2} \tag{9.3}$$

Substituting this in the Laplace equation, Eq. (9.1), we get

$$\frac{\Phi(x+\Delta_x, y) - 2\Phi(x,y) + \Phi(x-\Delta_x, y)}{\Delta_x^2} + \frac{\Phi(x, y+\Delta_y) - 2\Phi(x,y) + \Phi(x, y-\Delta_y)}{\Delta_y^2} = 0$$

$$\Rightarrow 2\Phi(x,y)\left[\frac{1}{\Delta_x^2}+\frac{1}{\Delta_y^2}\right]=\frac{\Phi(x+\Delta_x,y)+\Phi(x-\Delta_x,y)}{\Delta_x^2}+\frac{\Phi(x,y+\Delta_y)+\Phi(x,y-\Delta_y)}{\Delta_y^2}$$

If the grid-size in both the x- and y-directions are equal, $\Delta_x = \Delta_y = \Delta$, the above equation simplifies to

$$\Phi(x,y)=\frac{\Phi(x+\Delta,y)+\Phi(x-\Delta,y)+\Phi(x,y+\Delta)+\Phi(x,y-\Delta)}{4}$$

$$\Phi_{ij}=\frac{\Phi_{i+1,j}+\Phi_{i-1,j}+\Phi_{i,j+1}+\Phi_{i,j-1}}{4} \tag{9.4}$$

where Φ_{ij} is the scalar potential at the point, $(x_i = i\Delta, y_i = j\Delta)$. It can be seen from Eq. (9.4) that in the discretized version, the scalar potential at a point is simply given by the average of the potentials at the 4 neighbouring points. This also demonstrates the Earnshaw's theorem. If the potential at a point is the average of its neighbouring values, then it can neither be the maximum or minimum of these values but must be somewhere in between. This prevents any point in the region to be a local minima or maxima.

In order to solve the Laplace equation, we also need to know the boundary conditions since otherwise the solution of this equation is trivially zero. Consider a closed region of space where the potential has been specified on the boundary. We begin by forming a grid on the given space and initiating the potential, Φ, such that it is equal to the given values on the boundaries of this region and zero everywhere inside. Now, we scan all the grid coordinates in this region, and successively apply Eq. (9.4). In the first iteration, only points near the boundary get updated and all other grid points still have zero potential. We again scan all the grid points (except those at the boundary) and apply Eq. (9.4). Now more grid points get a non-zero value of the potential, and the points which had earlier had a non-zero value get updated. This process of applying Eq. (9.4) successively to all grid points goes on for a few rounds till we reach an equilibrium, i. e., when the potential values at the grid point no longer change by a significant amount. At each iteration, we evaluate the percentage change in the potential from the previous iteration and the iterations are stopped when this percentage change is lower than a certain threshold which depends on the accuracy desired,

$$\sum_{ij}\left|\Phi_{ij}^{(n+1)}-\Phi_{ij}^{(n)}\right| \le \epsilon \sum_{ij}\left|\Phi_{ij}^{(n)}\right|$$

where $\epsilon \ll 1$ is a small parameter.

9.2 Runge-Kutta Method

A very important aspect of any numerical technique is to evaluate the error that is incurred by the discretization process. No numerical scheme can be perfectly accurate, mainly because it replaces the limit of $\Delta \to 0$ with a finite value of Δ, which cannot be reduced beyond a certain limit due to finite computational resources. This error can be estimated through a Taylor expansion of the function to be integrated,

$$f(x + \Delta_x) = f(x) + \Delta_x f'(x) + \frac{\Delta_x^2}{2} f''(x) + \cdots \tag{9.5}$$

Thus, Eq. (9.2) is correct up to the first order since we have dropped terms of $\mathcal{O}(\Delta_x^2)$. This is very poor accuracy but still works for solving the Laplace equation since we get to perform multiple iterations to get to the actual solution. However, this luxury is not available in many problems of practical interest. In particular, problems which involve an evolution over time cannot be iterated recursively since the quantity to be integrated takes a forward march without any opportunity to re-evaluate the preceeding values. Hence, for such problems, it is often important to incorporate a few more terms from the Taylor series expansion while evaluating the derivatives. As we will see in later sections of this chapter, this problem does not arise for the numerical evaluation of Maxwell's equation in free space and Eq. (9.3) magically gives exact results for a certain specific choice of the discretization parameter, Δ. However, more care needs to be taken while solving for the dynamics of charged particles under the influence of electromagnetic fields.

One of the most common methods used to solve time-domain equations to a higher degree of accuracy is known as the *Runge-Kutta method.* We will discuss this method only up to the 4th order of accuracy (which means that the neglected terms are of $\mathcal{O}(\Delta^5)$), but higher order versions are also available and used in certain specific applications. For the sake of simplicity and without loss of generality, consider the 1D equation

$$\frac{dy}{dt} = f(y, t)$$

Using Eq. (9.5), we could write,

$$y(t + \Delta) = y(t) + \Delta y'(t) + \frac{\Delta^2}{2} y''(t) + \cdots$$

Neglecting terms of $\mathcal{O}(\Delta^2)$, we get

$$\begin{aligned} y_{n+1} &\approx y_n + \Delta y'(t_n) \\ &\approx y_n + \Delta f(y_n, t_n) \end{aligned} \tag{9.6}$$

where $t_{n+1} = t_n + \Delta$ and $y_n = y(t_n)$, which is known as the first-order Runge-Kutta solution.

In order to find an expression for the second-order Runge-Kutta method, we need to keep one more term in Eq. (9.6),

$$y_{n+1} \approx y_n + \Delta f(y_n, t_n) + \frac{\Delta^2}{2} y''(t_n) \tag{9.7}$$

for solving which we need to find an expression for $y''(t_n)$ in terms of $f(y_n, t_n)$ and y_n. Using chain rule, we have

$$\begin{aligned} y''(t) &= \frac{df}{dt} \\ &= \frac{\partial f}{\partial t} + \frac{dy}{dt}\frac{\partial f}{\partial y} \\ &= \frac{\partial f}{\partial t} + f\frac{\partial f}{\partial y} \end{aligned}$$

Substituting this in Eq. (9.7), we get

$$\begin{aligned} y_{n+1} &\approx y_n + \Delta f(y_n, t_n) + \frac{\Delta^2}{2}\left[\frac{\partial f}{\partial t} + f\frac{\partial f}{\partial y}\right]_{(y_n, t_n)} \\ &= y_n + \Delta\left[f(y_n, t_n) + \frac{\Delta}{2}\frac{\partial f}{\partial t}\bigg|_{(y_n, t_n)} + \frac{\Delta}{2} f(y_n, t_n)\frac{\partial f}{\partial y}\bigg|_{(y_n, t_n)}\right] \\ &\approx y_n + \Delta f\left(y_n + \frac{k_1}{2}, t_n + \frac{\Delta}{2}\right) \end{aligned} \tag{9.8}$$

where $k_1 = \Delta f(y_n, t_n)$, neglecting terms of $\mathcal{O}(\Delta^3)$. Thus, the second-order Runge-Kutta method proceeds in two steps. We first evaluate the function at the mid-point, $t_n^\star = t_n + \Delta/2$, and then use it to evaluate the value of y at $t = t_{n+1}$.

Adopting a similar approach leads to the fourth-order Runge-Kutta method, which gives,

$$\begin{aligned} k_1 &= \Delta f(y_n, t_n) \\ k_2 &= \Delta f\left(y_n + \frac{k_1}{2}, t_n + \frac{\Delta}{2}\right) \\ k_3 &= \Delta f\left(y_n + \frac{k_2}{2}, t_n + \frac{\Delta}{2}\right) \\ k_4 &= \Delta f(y_n + k_3, t_n + \Delta) \\ y_n + 1 &= y_n + \frac{1}{6}(k_1 + 2k_2 + 2k_3 + k_4) + \mathcal{O}\left(\Delta^5\right) \end{aligned} \tag{9.9}$$

Now, obviously a fourth-order method is likely to be more accurate than a second-order or first-order method, and this is indeed the case in most applications where the fourth-order expression provides sufficient accuracy and there is no need to go to higher order expressions. But, as mentioned above, in some specific cases like the Maxwell's even the first order can give perfectly accurate results. And of course, there can be specific applications where this fourth-order expression is not sufficiently accurate. In that case, there are three options which one can choose from. Firstly, one can write higher order expressions for the Runge-Kutta method and there are also commercially available libraries which provide solvers accurate up to the sixth order. Secondly, one can adopt other methods that are inherently more accurate, like the Bulirsch-Stoer or predictor-corrector methods. Thirdly, one can use the same fourth order expressions given by Eq. (9.9) but with adaptive step size. In this case, the value for y_{n+1} is evaluated using Eq. (9.9) for a particular value of step size, Δ, and an estimate for the error is also obtained. If the error is within a certain limit, this value is considered to be acceptable, otherwise the value of y_{n+1} is recalculated using a step size that is reduced by a factor of 2 or 3. This process is continued till the desired accuracy is reached or till the lower bound of the step size is reached. Here, it is important to note that y_{n+1} is not necessarily equal to $y(t_0 + n\Delta)$ since the step size used in each step of integration may be different.

9.3 Wave Equation: FDTD Method

In Section 9.1, we discussed the discretization procedure for numerically solving the Laplace equation. This equation gives the solution of the electrostatic potential in free space. We know that free space can also support plane electromagnetic waves, in which case, the equation of the electric potential is given by

$$\nabla^2 \Phi - \frac{1}{c^2}\frac{\partial^2 \Phi}{\partial t^2} = 0$$

where c is the speed of light. Like we did to discretize the Laplace equation, we can use the same Eq. (9.3) to solve the wave equation. This is known as the *Finite-Difference Time-Domain* (FDTD) method and is one of the most widely used numerical methods in computational electromagnetics. For the sake of simplicity, we again choose to work with the 1D version of Eq. (9.10),

$$\frac{\partial^2 \Phi}{\partial x^2} - \frac{1}{c^2}\frac{\partial^2 \Phi}{\partial t^2} = 0 \tag{9.10}$$

Using Eq. (9.3), we can write

$$\frac{\partial^2 \Phi}{\partial x^2} \approx \frac{\Phi(x_i + \Delta_x, t_n) - 2\Phi(x_i, t_n) + \Phi(x_i - \Delta_x, t_n)}{\Delta_x^2}$$

$$\frac{\partial^2 \Phi}{\partial t^2} \approx \frac{\Phi(x_i, t_n + \Delta_t) - 2\Phi(x_i, t_n) + \Phi(x_i, t_n - \Delta_t)}{\Delta_t^2}$$

Substituting this in Eq. (9.10), we get

$$\begin{aligned}
&\frac{\Phi(x_{i+1}, t_n) - 2\Phi(x_i, t_n) + \Phi(x_{i-1}, t_n)}{\Delta_x^2} \\
&\quad - \frac{\Phi(x_i, t_{n+1}) - 2\Phi(x_i, t_n) + \Phi(x_i, t_{n-1})}{c^2\Delta_t^2} \approx 0 \\
&\Rightarrow \frac{c^2\Delta_t^2}{\Delta_x^2}\left[\Phi_{i+1,n} - 2\Phi_{i,n} + \Phi_{i-1,n}\right] \\
&\qquad - \left[\Phi_{i,n+1} - 2\Phi_{i,n} + \Phi_{i,n-1}\right] \approx 0
\end{aligned}$$

$$\Rightarrow \Phi_{i,n+1} \approx \frac{c^2 \Delta_t^2}{\Delta_x^2} \left[\Phi_{i+1,n} - 2\Phi_{i,n} + \Phi_{i-1,n}\right] + 2\Phi_{i,n} - \Phi_{i,n-1} \tag{9.11}$$

If we choose $\Delta_x = c\Delta_t$, the above equation becomes,

$$\Rightarrow \Phi_{i,n+1} = \Phi_{i+1,n} + \Phi_{i-1,n} - \Phi_{i,n-1} \tag{9.12}$$

This particular choice of the step-size, $\Delta_t = \Delta_x/c$, is known as the *magic time-step* since in this case, Eq. (9.12) is an exact solution for the wave equation and not a mere approximation. This is very counterintuitive, since while deriving Eq. (9.3), we had to neglect the second order terms and so one would expect Eq. (9.12) to have the same order of accuracy. However, as we will now see, for this particular time-step, Eq. (9.12) is an exact solution of the wave equation! In order to see this, let us go back to the wave equation given by Eq. (9.10). The general solution to this equation is a superposition of two waves, one going in the forward direction and another going in the backward direction

$$\Phi(x, t) = \xi(x - ct) + \eta(x + ct) \tag{9.13}$$

where $\xi(\cdot)$ and $\eta(\cdot)$ are arbitrary functions of its variable. It can be easily verified that Eq. (9.13) is a valid solution by simply substituting it in Eq. (9.10). Now, if Eq. (9.12) is indeed the exact solution of the wave equation, then the resulting function, Φ, must satisfy the wave expression, Eq. (9.13), at all points of x for all time t. Let us assume that at a certain time $t = t_n$, the solution indeed satisfies the wave equation exactly. Hence, the solution for $t = t_{n+1}$ obtained using Eq. (9.12) must also satisfy Eq. (9.13). On substituting Eq. (9.13) on the RHS of Eq. (9.12), we get

$$\begin{aligned}
\Phi_{i,n+1} &= \Phi_{i+1,n} + \Phi_{i-1,n} - \Phi_{i,n-1} \\
&= [\xi(x_{i+1} - ct_n) + \eta(x_{i+1} + ct_n)] \\
&\quad + [\xi(x_{i-1} - ct_n) + \eta(x_{i-1} + ct_n)] \\
&\quad + [\xi(x_i - ct_{n-1}) + \eta(x_i + ct_{n-1})] \\
&= [\xi((i+1)\Delta_x - cn\Delta_t) + \eta((i+1)\Delta_x + cn\Delta_t)] \\
&\quad + [\xi((i-1)\Delta_x - cn\Delta_t) + \eta((i-1)\Delta_x + cn\Delta_t)] \\
&\quad + [\xi(i\Delta_x - c(n-1)\Delta_t) + \eta(i\Delta_x + c(n-1)\Delta_t)] \\
&= [\xi((i+1-n)\Delta_x) + \eta((i+1+n)\Delta_x)] \\
&\quad + [\xi((i-1-n)\Delta_x) + \eta((i-1+n)\Delta_x)] \\
&\quad + [\xi((i-n+1)\Delta_x) + \eta((i+n-1)\Delta_x)] \\
&= \xi((i-1-n)\Delta_x) + \eta((i+1+n)\Delta_x) \\
&= \xi(i\Delta_x - (n+1)\Delta_x) + \eta(i\Delta_x + (n+1)\Delta_x) \\
\Rightarrow \Phi_{i,n+1} &= \xi(x_i - ct_{n+1}) + \eta(x_i + ct_{n+1})
\end{aligned}$$

which is clearly an exact solution of Eq. (9.10). So if we start our numerical computation with the correct initial conditions, we will always be led to an exact solution of our wave equation despite the approximations made by discarding the higher order terms in the Taylor series expansion.

9.4 FDTD Dispersion Relation

We know from Eq. (1.18) that the dispersion relation of a plane wave in free space is given by $\omega^2 = c^2 k^2$ and this is indeed satisfied by the numerical solution given by Eq. (9.12), which is valid only when we choose the magic time-step, $\Delta_t = \Delta_x / c$. However, in some problems, it may become necessary to choose a different time-step and then the numerical solution given by Eq. (9.11) is no longer exact. In this case, it is important to evaluate the corresponding dispersion relation and analyse its properties. In order to do this, let us substitute the plane wave solution, $\Phi_{i,n} = \exp\left[j\left(\tilde{k}x_i - \omega t_n\right)\right]$, where $j = \sqrt{-1}$, $\tilde{k} = k_{real} + jk_{imag}$ is the complex wave number and ω is the angular frequency, into Eq. (9.11),

$$\begin{aligned}
e^{[j(\tilde{k}x_i - \omega t_{n+1})]} &\approx \frac{c^2\Delta_t^2}{\Delta_x^2}\left[e^{[j(\tilde{k}x_{i+1} - \omega t_n)]} - 2e^{[j(\tilde{k}x_i - \omega t_n)]}\right.\\
&\quad \left. + e^{[j(\tilde{k}x_{i-1} - \omega t_n)]}\right] + 2e^{[j(\tilde{k}x_i - \omega t_n)]} - e^{[j(\tilde{k}x_i - \omega t_{n-1})]}\\
\Rightarrow e^{[j(\tilde{k}x_i - \omega t_n - \omega\Delta_t)]} &\approx \frac{c^2\Delta_t^2}{\Delta_x^2}\left[e^{[j(\tilde{k}x_i + \tilde{k}\Delta_x - \omega t_n)]} - 2e^{[j(\tilde{k}x_i - \omega t_n)]}\right.\\
&\quad \left. + e^{[j(\tilde{k}x_i + \tilde{k}\Delta_x - \omega t_n)]}\right] + 2e^{[j(\tilde{k}x_i - \omega t_n)]} - e^{[j(\tilde{k}x_i - \omega t_n + \omega\Delta_t)]}\\
\Rightarrow e^{-j\omega\Delta_t} &\approx \frac{c^2\Delta_t^2}{\Delta_x^2}\left[e^{j\tilde{k}\Delta_x} - 2 + e^{j\tilde{k}\Delta_x}\right] + 2 - e^{j\omega\Delta_t}\\
\Rightarrow \frac{e^{j\omega\Delta_t} + e^{-j\omega\Delta_t}}{2} &\approx \frac{c^2\Delta_t^2}{\Delta_x^2}\left[\frac{e^{j\tilde{k}\Delta_x} + e^{j\tilde{k}\Delta_x}}{2} - 1\right] + 1\\
\Rightarrow \cos(\omega\Delta_t) &\approx \frac{c^2\Delta_t^2}{\Delta_x^2}\left[\cos\left(\tilde{k}\Delta_x\right) - 1\right] + 1
\end{aligned}$$

which implies that, for a given angular frequency, the complex wave number is given by

$$\begin{aligned}
\tilde{k} &= \frac{1}{\Delta_x}\cos^{-1}\left\{1 - \frac{\Delta_x^2}{c^2\Delta_t^2}\left[1 - \cos(\omega\Delta_t)\right]\right\} \\
&= \frac{1}{\Delta_x}\cos^{-1}\left\{1 - \frac{1}{S^2}\left[1 - \cos\frac{2\pi S}{N_\lambda}\right]\right\}
\end{aligned} \tag{9.14}$$

which is the required dispersion relation, where $S = c\Delta_t / \Delta_x$ is the numerical stability factor (or Courant number) and $N_\lambda = \lambda / \Delta_x$ is the grid sampling resolution.

Substituting $\Delta_x = c\Delta_t$ into Eq. (9.14) gives $\tilde{k} = \pm\omega/c$ as expected. We also expect $\tilde{k} = \omega/c$ to hold when the spatial and temporal step-size become very small, i. e., $\Delta_x, \Delta_t \to 0$. In order to verify this, let us take the limit of Eq. (9.14),

$$\begin{aligned}
\tilde{k}_{SS} &= \lim_{\Delta_x,\Delta_t\to 0} \frac{1}{\Delta_x} \cos^{-1}\left\{1 - \frac{\Delta_x^2}{c^2\Delta_t^2}\left[1 - \cos(\omega\Delta_t)\right]\right\} \\
&= \lim_{\Delta_x,\Delta_t\to 0} \frac{1}{\Delta_x} \cos^{-1}\left\{1 - \frac{\Delta_x^2}{c^2\Delta_t^2}\left[1 - \left\{1 - \frac{\omega^2\Delta_t^2}{2}\right\}\right]\right\} \\
\tilde{k}_{SS} &= \lim_{\Delta_x,\Delta_t\to 0} \frac{1}{\Delta_x} \cos^{-1}\left\{1 - \frac{\omega^2\Delta_x^2}{2c^2}\right\} \\
&= \lim_{\Delta_x,\Delta_t\to 0} \frac{1}{\Delta_x} \left\{\sqrt{2}\sqrt{\frac{\omega^2\Delta_x^2}{2c^2}} + \frac{1}{6\sqrt{2}}\left(\frac{\omega^2\Delta_x^2}{2c^2}\right)^{3/2} + \cdots\right\} \\
&= \pm\frac{\omega}{c}
\end{aligned}$$

Except for these two special conditions, magic time-step and vanishing step-size, numerical wave propagation will always be dispersive. This implies that, even in free space, numerically obtained the phase and group velocity will always depend on the frequency. From Eq. (9.14), we can see that if $\cos(\omega\Delta_t) < 0$, the wave number may even be complex leading to artificial losses. For the solution of Eq. (9.14) to be real, we must have

$$\begin{aligned}
1 - \frac{1}{S^2}\left[1 - \cos\frac{2\pi S}{N_\lambda}\right] &> -1 \\
\Rightarrow 1 - \cos\frac{2\pi S}{N_\lambda} &< 2S^2 \\
\Rightarrow \cos\frac{2\pi S}{N_\lambda} &> 1 - 2S^2 \\
\Rightarrow \frac{2\pi S}{N_\lambda} &< \cos^{-1}\left(1 - 2S^2\right) \\
\Rightarrow N_\lambda &> \frac{2\pi S}{\cos^{-1}\left(1 - 2S^2\right)} = N_{\lambda,\text{threshold}}
\end{aligned} \tag{9.15}$$

For Eq. (9.15) to be meaningful, we also must have $-1 \le 1 - 2S^2 \le 1$, which implies that

$$S \le 1 \tag{9.16}$$

It can shown that if these two conditions, Eqs. (9.15) and (9.16), are not satisfied, then the phase velocity can also exceed the speed of light and there can be artificial numerical instabilities. Hence, for numerical simulation of the wave equation, it is important to ensure that these conditions are satisfied.

9.5 Dispersive Materials

The FDTD method described above can be used to solve for the propagation of electromagnetic waves in free space and non-dispersive media (where the permittivity and permeability do not depend on frequency). However, as we have seen in previous chapters, both plasma waves and surface plasmons are inherently dispersive. In this section, we will discuss how to numerically solve the Maxwell's equations for materials where the permittivity depends on frequency. As we saw in Eqs. (2.5) and (1.61), the permittivity of a general dispersive medium is given by

$$\begin{aligned}\epsilon(\omega) &= \epsilon_0 + \frac{Ne^2}{M}\sum_i \frac{f_i}{\omega_i^2 - \omega^2 - j\omega\gamma_i}\\ &= \epsilon_0 + \epsilon_0\chi(\omega)\end{aligned}$$

and the relation between the electric displacement vector and electric field is given by

$$\begin{aligned}\vec{D}(\omega) &= \epsilon(\omega)\vec{E}(\omega)\\ \Rightarrow \vec{D}(t) &= \epsilon(t) \star \vec{E}(t)\\ &= \int_0^{\infty} \vec{E}\left(t - t'\right)\epsilon\left(t'\right)dt'\\ &= \epsilon_0\vec{E}(t) + \epsilon_0\int_0^t \vec{E}\left(t - t'\right)\chi\left(t'\right)dt' \end{aligned} \tag{9.17}$$

where $\star$ represents the convolution operator, the lower limit of the integration is at 0 since we are dealing with a causal system (see Eq. (1.53)) and the upper limit is t since the electric field, $\vec{E}(t)$, is assumed to be zero for $t < 0$. Using the frequency-domain equation for our analysis would certainly be easier since it involves a simple multiplication, but in plasma science and plasmonics, it is of primary interest to simulate the time evolution of the field quantities. Hence, it is imperative to directly solve the time-domain equation although it is more complicated. There are other numerical methods like the Finite Element Method (FEM) which solve the frequency-domain equations, but they have several limitations and are not very widely used for these problems.

For numerical computation of the electric and magnetic fields for dispersive materials and other problems with sources, it has shown to be more accurate to solve the individual Maxwell's solution instead of directly solving the wave equation. Discretizing Eq. (9.17), we get

$$\Rightarrow \vec{D}_{i,n} = \epsilon_0\vec{E}_{i,n} + \epsilon_0\int_{t'=0}^{n\Delta_t} \vec{E}_i\left(n\Delta_t - t'\right)\chi\left(t'\right)dt' \tag{9.18}$$

where the subscript i refers to the spatial grid point and n refers to the temporal grid point. The next step is to find an expression for $\vec{E}_i\left(n\Delta_t - t'\right)$ in terms of the electric field values at the grid points. This can be done by choosing a variable m such that

$t'' = t' - m\Delta_t < \Delta_t$, which gives

$$\begin{aligned}\vec{E}_i\left(n\Delta_t - t'\right) &= \vec{E}_i\left(n\Delta_t - m\Delta_t - t''\right) \\ &= \vec{E}_{i,n-m} - \frac{\vec{E}_{i,n-m} - \vec{E}_{i,n-m-1}}{\Delta_t} t''\end{aligned} \tag{9.19}$$

Substituting this in Eq. (9.18), we get

$$\begin{aligned}\Rightarrow \vec{D}_{i,n} &= \epsilon_0 \vec{E}_{i,n} + \epsilon_0 \int_{t'=0}^{n\Delta_t} \left[\vec{E}_{i,n-m} - \frac{\vec{E}_{i,n-m} - \vec{E}_{i,n-m-1}}{\Delta_t} t''\right] \chi\left(t'\right) dt' \\ &= \epsilon_0 \vec{E}_{i,n} + \epsilon_0 \sum_{m=0}^{n-1} \vec{E}_{i,n-m} \int_{m\Delta_t}^{(m+1)\Delta_t} \chi\left(t'\right) dt' \\ &\quad - \epsilon_0 \sum_{m=0}^{n-1} \frac{\vec{E}_{i,n-m} - \vec{E}_{i,n-m-1}}{\Delta_t} \int_{m\Delta_t}^{(m+1)\Delta_t} \chi\left(t'\right)\left(t' - m\Delta_t\right) dt' \\ &= \epsilon_0 \vec{E}_{i,n} + \epsilon_0 \sum_{m=0}^{n-1} \left[\vec{E}_{i,n-m}\chi_m - \left(\vec{E}_{i,n-m} - \vec{E}_{i,n-m-1}\right)\xi_m\right]\end{aligned} \tag{9.20}$$

where

$$\begin{aligned}\chi_m &= \int_{m\Delta_t}^{(m+1)\Delta_t} \chi\left(t'\right) dt' \\ \xi_m &= \frac{1}{\Delta_t} \int_{m\Delta_t}^{(m+1)\Delta_t} \chi\left(t'\right)\left(t' - m\Delta_t\right) dt'\end{aligned}$$

Now we need to discretize the Ampere's law (for current free region) so as to get the expression for time-evolution of the electric field,

$$\begin{aligned}\vec{\nabla} \times \vec{H} &= \frac{\partial \vec{D}}{\partial t} \\ &= \frac{\vec{D}_{i,n+1} - \vec{D}_{i,n}}{\Delta_t} \\ \Rightarrow \vec{D}_{i,n+1} &= \vec{D}_{i,n} + \Delta_t\left(\vec{\nabla} \times \vec{H}\right)\end{aligned} \tag{9.21}$$

Substituting the discretized version of the electric displacement vector, $\vec{D}$, from Eq. (9.20) into Eq. (9.21), we get

$$\begin{aligned}\Rightarrow \vec{E}_{i,n+1} &= -\sum_{m=0}^{n} \left[\vec{E}_{i,n+1-m}\chi_m - \left(\vec{E}_{i,n+1-m} - \vec{E}_{i,n-m}\right)\xi_m\right] \\ &\quad + \vec{E}_{i,n} + \sum_{m=0}^{n-1} \left[\vec{E}_{i,n-m}\chi_m - \left(\vec{E}_{i,n-m} - \vec{E}_{i,n-m-1}\right)\xi_m\right] \\ &\quad + \frac{\Delta_t}{\epsilon_0}\left(\vec{\nabla} \times \vec{H}\right)\end{aligned}$$

$$\begin{aligned}\Rightarrow \vec{E}_{i,n+1} &= -\vec{E}_{i,n+1}\chi_0 + \vec{E}_{i,n+1}\xi_0 - \vec{E}_{i,n}\xi_0 \\ &\quad - \sum_{m=1}^{n}\left[\vec{E}_{i,n+1-m}\chi_m - \left(\vec{E}_{i,n+1-m} - \vec{E}_{i,n-m}\right)\xi_m\right] \\ &\quad + \vec{E}_{i,n} + \sum_{m=0}^{n-1}\left[\vec{E}_{i,n-m}\chi_m - \left(\vec{E}_{i,n-m} - \vec{E}_{i,n-m-1}\right)\xi_m\right] \\ &\quad + \frac{\Delta_t}{\epsilon_0}\left(\vec{\nabla}\times\vec{H}\right)\end{aligned}$$

$$\begin{aligned}&\Rightarrow (1+\chi_0-\xi_0)\,\vec{E}_{i,n+1} \\ &\quad = -\sum_{m=0}^{n-1}\left[\vec{E}_{i,n-m}\chi_{m+1} - \left(\vec{E}_{i,n-m} - \vec{E}_{i,n-m-1}\right)\xi_{m+1}\right] \\ &\qquad + (1-\xi_0)\,\vec{E}_{i,n} + \sum_{m=0}^{n-1}\left[\vec{E}_{i,n-m}\chi_m - \left(\vec{E}_{i,n-m} - \vec{E}_{i,n-m-1}\right)\xi_m\right] \\ &\qquad + \frac{\Delta_t}{\epsilon_0}\left(\vec{\nabla}\times\vec{H}\right)\end{aligned}$$

which finally gives the expression for the time evolution of the electric field,

$$\begin{aligned}\Rightarrow \vec{E}_{i,n+1} &= \frac{(1-\xi_0)}{(1+\chi_0-\xi_0)}\vec{E}_{i,n} + \frac{\Delta_t}{\epsilon_0\,(1+\chi_0-\xi_0)}\left(\vec{\nabla}\times\vec{H}\right) \\ &\quad + \frac{1}{(1+\chi_0-\xi_0)}\sum_{m=0}^{n-1}\vec{E}_{i,n-m}\,(\chi_m - \chi_{m+1}) \\ &\quad - \frac{1}{(1+\chi_0-\xi_0)}\sum_{m=0}^{n-1}\left(\vec{E}_{i,n-m} - \vec{E}_{i,n-m-1}\right)(\xi_m - \xi_{m+1})\end{aligned}$$

9.6 FDTD Boundary Conditions

Though we have briefly described the FDTD method for numerical computation of Maxwell's equations, the details of this method are seldom used by practitioners to do the programming. This is because several commercial and free softwares (e. g., MEEP by MIT) are available which does these simulations. In many practical situations, the size of the computational domain is too large and serial processing takes too long to compute meaningful results. Hence, it often becomes necessary to use some kind of parallelisation either through multiple CPUs or GPUs (e. g., XFDTD by Remcom, which was used to do the spoof surface plasmon simulations discussed in Chapter 6). However, in order to efficiently use these softwares, it is important to have a basic understanding of the FDTD method which helps in properly setting the parameters (e. g., N_λ and S given by Eqs. (9.15) and (9.16)).

Another important concept one needs to understand (even for using these software tools) is the boundary condition to be used while computationally analysing a particular EM scenario. This concept of boundary conditions is different from what we encountered in Eq. (1.33), which are the conditions that fields across the interface between two different media must satisfy. However, in the case of computational electromagnetics, there is an additional concern since all computational grids (see Figure 9.1) must be finite. We can analytically solve the Maxwell's equations over the entire infinite 2D or 3D space, but for doing computations, we must specify a certain finite region. Hence, we must specify the conditions that the electric field must satisfy at these boundaries of the computational domain. There are, of course, many different possibilities available but in simulations of plasmas and plasmonic waves, the two major boundary conditions used are: *Periodic Boundary Conditions* (PBC) and *Perfectly Matched Layer* (PML).

As the name suggests, a PBC is used when the problem being analysed contains periodic structures. One example could be the spoof surface plasmon discussed in Chapter 6, or a periodic array of nanoparticles leading to Localised Surface Plasmon Resonance discussed in Section 5.4. One way to computationally solve these problems could be to create a grid over the entire 2D array of holes or nanoparticles. But this would be computationally very expensive. What a PBC does is that it incorporates the effects of this periodicity into the discretizated field equations itself. So, for studying an array of 2D nanoparticles, it is sufficient to have just one nanoparticle in a computational domain of appropriate size and then apply PBC.

Though PBC is a very useful technique, one of its limitations is that if the problem under consideration has finite sources (charges/current) at a certain point in the physical domain, then these sources are also periodically replicated in all the computational cells. In this case, a PBC is not suitable and there is no option but to simulate the whole physical domain with all the nanoparticles/holes explicitly put in. In this case also, we do need an appropriate boundary condition since the computational domain must be finite. One of the usual requirements in such problems is that the EM field going out from the region under consideration must not reflect back from the boundary of the computational domain. This is achieved by applying what is known as the Perfectly Matched Layer (PML) boundary condition.

A PML essentially consists of lossy or absorbing media placed at the boundary of the computational domain. This lossy medium is also anisotropic since only the field components travelling normally into the layer must be absorbed and not the field components travelling parallel to this layer. The word *matched* comes from the study of *transmission lines*, which are just simplified mathematical models for analytical study of wave propagation in waveguides discussed in Section 1.8. In the transmission line model, every waveguide has a characteristic impedance usually denoted by Z_0. If the transmission line is terminated using a load impedance, Z_L, then it can be shown that no EM energy will be reflected back if $Z_L = Z_0$, in which case the transmission line is said to be matched. However, in the case of PML, though the terminology is the same, the implementation procedure is different. Unlike transmission lines where the matching impedance, Z_L, can be purely real and therefore lossless, a PML works using lossy medium. One of the reasons for this difference is also that a problem in computational electromagnetics usually does not have a characteristic impedance, Z_0, which can be equated to the load impedance. Electromagnetic fields of various kinds can exist within a computational domain and can impinge on the boundary from infinitely many directions. Due to this immense variation, the best option is to use a lossy material in order to ensure minimum reflection at the boundaries. A PML implemented in practical computational algorithms usually consists of multiple layers with gradually increasing conductivity (which causes losses), which helps in further reducing reflections.

Appendix: Legendre Polynomials

The Laplace equation in spherical polar coordinates can be written as

$$\frac{1}{r}\frac{\partial^2}{\partial r^2}(r\Phi) + \frac{1}{r^2 \sin\theta}\frac{\partial}{\partial\theta}\left(\sin\theta\frac{\partial\Phi}{\partial\theta}\right) + \frac{1}{r^2\sin^2\theta}\frac{\partial^2\Phi}{\partial\phi^2} = 0$$

If a product form is assumed,

$$\Phi(r,\theta,\phi) = \frac{U(r)}{r}P(\theta)\,Q(\phi)$$

we get

$$r^2\sin^2\theta\left[\frac{1}{U}\frac{d^2U}{dr^2} + \frac{1}{Pr^2\sin\theta}\frac{d}{d\theta}\left(\sin\theta\frac{dP}{d\theta}\right)\right] + \frac{1}{Q}\frac{d^2Q}{d\phi^2} = 0$$

Since the dependence on ϕ has been isolated, we can write

$$\frac{1}{Q}\frac{d^2Q}{d\phi^2} = -m^2$$

$$\Rightarrow Q = e^{\pm im\phi}$$

where m must be an integer since Q must be periodic with period 2π. Similarly, we can write

$$\frac{1}{\sin\theta}\frac{d}{d\theta}\left(\sin\theta\frac{dP}{d\theta}\right) + \left[l(l+1) - \frac{m^2}{\sin\theta}\right]P = 0$$

$$\frac{d^2U}{dr^2} - \frac{l(l+1)}{r^2}U = 0$$

where l is another constant. The solution for U can be written as

$$U = Ar^{l+1} + Br^{-l}$$

In the equation for P, if we use the variable $x = \cos\theta$, we get

$$\frac{d}{dx}\left[\left(1-x^2\right)\frac{dP}{dx}\right] + \left[l(l+1) - \frac{m^2}{1-x^2}\right]P = 0$$

which is the *generalized Legendre equation* and its solutions are known as the *associated Legendre polynomials*. If the system possesses azimuthal symmetry, we have $m = 0$ and the above is known as the *ordinary Legendre equation* and its solutions are the *Legendre polynomials* of order l where l must be zero or a positive integer for the series solution to converge,

$$P_0(x) = 1$$

$$P_1(x) = x$$

$$P_2(x) = \frac{1}{2}\left(3x^2 - 1\right)$$

$$P_3(x) = \frac{1}{2}\left(5x^3 - 3x\right)$$

$$P_4(x) = \frac{1}{8}\left(35x^4 - 30x^2 + 3\right)$$

https://doi.org/10.1515/9783110570038-163

A compact representation of the Legendre polynomials is given by the *Rodrigue's formula*

$$P_l(x) = \frac{1}{2^l l!} \frac{d^l}{dx^l} \left(x^2 - 1\right)^l$$

The functions $P_l(x)$ form a complete orthonormal set of functions on the interval $-1 \le x \le 1$.

Bibliography

[1] S. Chandrasekhar, "Stochastic Problems in Physics and Astronomy", *Rev. Mod. Phys.* 15, 1 (1943).

[2] D. R. Nicholson, *Introduction to Plasma Theory* (Wiley, New York, 1983).

[3] I. B. Bernstein, J. M. Greene, and M. D. Kruskal, "Exact Nonlinear Plasma Oscillations", *Phys. Rev.* 108, 546 (1957).

[4] K. Shah and H. S. Ramachandran, "Analytic, Nonlinearly Exact Solutions for an RF Confined Plasma", *Phys. Plasmas* 15, 062303 (2008).

[5] K. Shah and H. S. Ramachandran, "Space Charge Effects in RF Traps: Ponderomotive Concept and Stroboscopic Analysis", *Phys. Plasmas* 16, 062307 (2009).

[6] K. Shah, *Plasma Dynamics in Paul Traps*, Ph.D. dissertation, Indian Institute of Technology Madras, 2010.

[7] R. C. Davidson, *Methods in Nonlinear Plasma Theory* (Academic Press, New York, 1972).

[8] A. J. Lichtenberg and M. A. Lieberman, *Regular and Chaotic Dynamics* (Springer, New York, 1992).

[9] N. W. McLachlan, *Theory and Applications of Mathieu Functions* (Oxford University Press, Oxford, 1947).

[10] D. Maystre, "Analytic Properties of Diffraction Gratings", ch.2 in *Gratings: Theory and Numeric Applications*, ed. E. Popov (Institut Fresnel, CNRS, AMU 2012).

[11] J. Leuthold *et al.*, "Plasmonic Communications: Light on a Wire", *Optics and Photonics News* 24, 28 (2013).

[12] A. G. Brolo, "Plasmonics for future biosensors", *Nature Photonics* 6, 709 (2012).

[13] H. A. Atwater and A. Polman, "Plasmonics for improved photovoltaic devices", *Nature Materials* 9, 205 (2010).

[14] S. Baldelli, "Sensing: Infrared image upconversion", *Nature Photonics* 5, 75 (2011).

[15] J. B. Pendry, A. J. Holden, W. J. Stewart, and I. Youngs, "Extremely Low Frequency Plasmons in Metallic Mesostructures", *Physical Review Letters* 76, 4773 (1996).

[16] J. B. Pendry, L. Martin-Moreno, and F. J. Garcia-Vidal, "Mimicking Surface Plasmons with Structured Surfaces", *Science* 305, 847 (2004).

[17] S. H. Kim, T. T. Kim, S. Oh, J. Kim, H. Y. Park, and C. S. Kee, "Experimental Demonstration of Self-collimation of Spoof Surface Plasmons", *Phys. Rev. B* 83, 165109 (2011).

[18] S. Bhattacharya and K. Shah, "Multimodal Propagation of the Electromagnetic Wave on a Structured Perfect Electric Conductor (PEC) Surface", *Optics Communications* 328, 102 (2014).

[19] S. Bhattacharya and K. Shah, "Dispersion Relation and Self-collimation Frequency of Spoof Surface Plasmon Using Tight Binding Model", *Journal of Optics* 17, 065102 (2015).

[20] S. Bhattacharya, *Study of Electromagnetic Waves on a Structured Perfect Electric Conductor (PEC) Surface*, PhD dissertation, Indian Institute of Technology Delhi, 2016.

[21] J. C. Slater and G. F. Koster, "Simplified LCAO Method for the Periodic Potential Problem", *Phys. Rev. B* 94, 1498 (1954).

[22] P. O. Lowdin, "On the Non-orthogonality Problem Connected with the use of Atomic Wave Functions in the Theory of Molecules and Crystals", *The J. of Chem. Phys.* 18, 365 (1950).

[23] H. Goldstein, C. Poole and J. Safko, *Classical Mechanics* (Pearson, New York, 2011).

https://doi.org/10.1515/9783110570038-165

[24] V. G. Veselago, "The Electrodynamics of Substances with Simultaneously Negative Values of ϵ and μ", *Soviet Physics Uspekhi* 10, 509 (1968).

[25] W. Paul, "Electromagnetic Traps for Charged and Neutral Particles", *Rev. Mod. Phys.* 62, 531 (1990).

[26] A. Ashkin, J. M. Dziedzic, J. E. Bjorkholm and Steven Chu, "Observation of a Single-Beam Gradient Force Optical Trap for Dielectric Particles", *Optics Letters* 11, 288 (1986).

[27] K. C. Neuman and S. M. Block, "Optical Trapping", *Rev. Sci. Instrum.* 75, 2787 (2004).

[28] G. Volpe, R. Quidant, G. Badenes and Dmitri Petrov, "Surface Plasmon Radiation Forces", *Phys. Rev. Lett.* 96, 238101 (2006).

[29] M. L. Juan, M. Righini and R. Quidant, "Plasmon Nano-Optical Tweezers", *Nature Photonics* 5, 349 (2011).

[30] S. H. Strogatz, *Nonlinear Dynamics And Chaos* (Westview Press, Massachusetts, 2001).

[31] A. Taflove and S. C. Hagness, *Computational Electrodynamics: The Finite-Difference Time-Domain Method* (Artech House, London, 2005).

Index

Adiabatic invariance 42
Ampere-Maxwell Law 6
Ampere's circuital Law 5
Anomalous dispersion 27
Attractor 129
Autonomous system 135
Azimuthal angle 3, 102

Backward media 85
Backward propagation 140
Bound charge 9
Bound magnetic current 10
Boundary conditions 13, 141
Brewster's angle 15

Cartesian Coordinate System 100
Causality 120
Characteristic equation 130
Charge density 3, 11, 119
Collision 27, 33
Continuity equation 6, 46
Convolution 12, 76, 119, 148
Coordinate System 92, 103, 115
Coulomb's Law 2
Critical angle 15
Curl 105, 109
Current density 5, 11,
Cylindrical Coordinate System 102, 103

Decay constant 29, 71, 73
Degenerate node 130
Diamagnetism 10
Dielectric 9, 27, 55, 74
Dielectric gratings 63
Diffraction 78
Dipole surface plasmon 68
Dirac delta function 27, 119
Dispersion relation 146
Displacement current 6, 11
Distribution function 33, 45, 46
Divergence 105, 137
Divergence theorem 109, 138
Double negative metamaterials 83
Drude model 28

Earnshaw's theorem 90, 138
Effective permittivity 28
Eigenfrequency 80
Electric field 9, 22, 36, 128
Electric flux 3
Electric polarization current 10
Equifrequency contour 18, 19, 80
Evanescent wave 13, 15, 23, 76

Faraday's Law 4, 91
Ferroelectric 9
Ferromagnetism 10
FDTD method 144, 150
Fixed point 125–129
Floquet theorem 93
Fluorescence 89
Forward propagation 140
Free charge 10–11, 55, 70
Free current 10–11, 58
Frohlich condition 68, 90

Galilean transformation 113–115
Gauss' Law 2, 3
Gibbs-Boltzmann distribution 36, 42
Gradient force 95, 97
Gradient operator 105
Gradient theorem 109
Green's function 117, 120
Group velocity 17, 18

Hamiltonian 134
Helmholtz equation 56
Hopping parameters 79

Impulse response 119
Infrared photography 70

Jacobian matrix 127

Kretschmann configuration 63

Landau damping 37, 39, 49
Langmuir waves 48
Laplace equation 139
Legendre polynomial 153
Limit cycle 133
Lindstedt-Poincare method 94
Linear response theory 34, 37
Linearity 118

https://doi.org/10.1515/9783110570038-167

Linearization 38, 49
Liouville equation 34
LTI system 117
Localized surface plasmon resonance 65, 68
Lorentz force 34, 95
Lorentz transformation 113, 115
Lyapunov stable 129

Magic time-step 145
Magnetic flux 4
Mathieu's equation 92
Mie scattering 67, 89
Momentum equation 47

Near field excitation 65
Negative group velocity 85
Negative index material 84
Neutral fixed point 127

Ohm's law 28
Optical tweezer 95
Otto configuration 63

Paraelectric 9
Paramagnetism 10
Paul trap 90, 94
Perfect electric conductor 17, 74
Perfect lens 88
Perfectly matched layer 151
Periodic boundary conditions 151
Periodic lattice 79
Permeability 5, 11
Permittivity 11, 27, 50
Phase velocity 17
Plasma distribution 33
Plasma frequency 28
Plasma waves 34, 37
Plasmonic tweezer 97
Polar angle 103
Polarization 9–10
Poisson's equation 7, 29
Ponderomotive theory 39, 41
Position vector 100
Positive index materials 83
Prism coupling 63, 65
Propagation constant 13, 29
Propagation vector 14

Quasi-neutral plasma 38

Raman effect 89
Raman spectrum 89
Rayleigh scattering 89
Rayleigh-Sommerfeld formulation 76
Runge-Kutta method 141–142

Saddle node 127
Scalar 99, 107, 111
Scalar potential 105, 107
Self-collimation 77
Skin effect 16, 29
Snell's law 14, 85
Solar cells 170
Spherical Coordinate System 103, 108
Split ring resonators 83
Spoof surface plasmon 71
Stable fixed point 127–129
Stable limit cycle 133
Stability 120, 128
Star node 130
Stokes scattering 89
Superlens 86
Surface Enhanced Raman Scattering 88, 90
Surface plasmon ix, 16, 59, 97
Susceptibility 11

TE Modes 76
TM Modes 60
Tight binding model 79
Time invariance 118
Total internal reflection 15, 63
Transfer function 121–123

Unstable fixed point 127
Unstable limit cycle 133

Vector 99, 104, 116
Vector potential 111
Vlasov equation 33–35

Wave number 37, 147
Waveguide 19–21
Wire mesh 72, 83

www.ingramcontent.com/pod-product-compliance
Lightning Source LLC
Chambersburg PA
CBHW081534220326
41598CB00036B/6434
9783110569940